Condensed Matter Researches in Cryospheric Science

Condensed Matter Researches in Cryospheric Science

Special Issue Editors

Augusto Marcelli
Valter Maggi
Cunde Xiao

MDPI • Basel • Beijing • Wuhan • Barcelona • Belgrade

Special Issue Editors

Augusto Marcelli
Istituto Nazionale di Fisica Nucleare
Italy

Valter Maggi
University of Milano Bicocca
Italy

Cunde Xiao
Beijing Normal University
China

Editorial Office
MDPI
St. Alban-Anlage 66
4052 Basel, Switzerland

This is a reprint of articles from the Special Issue published online in the open access journal *Condensed Matter* (ISSN 2410-3896) from 2018 to 2019 (available at: https://www.mdpi.com/journal/condensedmatter/special_issues/cryospheric_science).

For citation purposes, cite each article independently as indicated on the article page online and as indicated below:

LastName, A.A.; LastName, B.B.; LastName, C.C. Article Title. *Journal Name* **Year**, *Article Number*, Page Range.

ISBN 978-3-03921-323-8 (Pbk)
ISBN 978-3-03921-324-5 (PDF)

Cover image: Dosegu' glacier from Passo Gavia, Valtellina (Italy). Courtesy by Stefano Pignotti.

Contents

About the Special Issue Editors

Valter Maggi, Prof. and Ph.D. Dr. Maggi's main fields of research are related to the characterization of atmospheric dust, mainly in Antarctica, Greenland, and Alpine areas, for long-term paleoclimatic influences of mineral phases on the atmosphere. He has participated in the main European ice core drilling projects for the reconstruction of past climatic changes and the environmental changes in Antarctica, Greenland, and the European Alps area. In 2008, he was a winner of the EU Descartes Prize, along with other 10 European colleagues, for collaborative research.

Cunde Xiao, Dr. and Prof. Dr. Xiao's major research focus has been ice core studies relating to the paleoclimate and paleoenvironment, and present-day cold-region meteorological and glaciological processes that impact environmental and climatic changes, cryospheric services, and values that link the cryosphere with the socioeconomy.

Augusto Marcelli, Dr. and Prof. Since 1984, Dr. Marcelli has been involved in synchrotron radiation research, and his main research areas include correlation phenomena in X-ray absorption spectroscopy, circular magnetic X-ray dichroism investigations, X-ray absorption in elements of geophysical interest, dust and aerosol characterization, and ultratrace detection for indoor and outdoor environmental research. Since 2001, he has collaborated with the Institute of High Energy Physics in Beijing and has been Visiting Professor at the University of Science and Technology at Hefei. In 2018, he was appointed as a scientific expert on bilateral policies and activities for the internationalization of scientific and technological research of the Italian Ministry of Foreign Affairs.

Preface to "Condensed Matter Researches in Cryospheric Science"

The aim of this thematic Special Issue, which collects a variety of papers published during the year 2018, is to display recent results obtained on the most widespread solid material occurring on the surface of the Earth—ice. Unfortunately, this material is dramatically decreasing in quantity and extent across the globe. The Antarctic icecap, which for millions of years was the largest solid mass of ice in the world and had recently expanded by binding some marginal, loose sea-supported ice masses, is not only decreasing in thickness, but is also splitting away all around its rim, liberating pollutants of various kinds into the Circum-Antarctic stream. The Arctic ice shelf, which since at least the end of the last glaciation has floated over the North Pole while the huge and complex kilometer-thick ice cover over Greenland was being formed, is melting away and getting thinner and thinner, to the point of letting commercial ships, which are no longer icebreakers, cross from the Barents Sea to the Bering Sea, while icebergs slide down from Greenland's frozen valleys into the upper Atlantic Ocean and gradually become submerged. The third significant ice cap, the Patagonian one, is getting considerably smaller while releasing fresh waters into the Argentinian lakes and into the Chilean rivers flowing into Pacific Ocean fjords, thus adding pollutants of undisputed terrestrial origin to the sea stream that borders the entire South America continent and discharging them in the tropical zone. The worst condition is that of the mountain glaciers scattered throughout the continents: they are not only decreasing in size so dramatically as to disappear altogether, but with their water discharge, they change the landscape and local environment of densely inhabited areas such as the Alps, the Caucasus, the Himalaya, the Andes, and even the Kilimanjaro, thus forcing people into mass migration. Even when local people's reactions to such environmental change are disregarded as a local problem, there is a hidden, major risk to be taken into consideration: intense ice melting on the major mountain barriers all over the word changes the orientation of the trade winds, and this modifies the regular sequence of rains, as has already happened in many parts of the globe where hurricanes have replaced the strong but calm monsoon rainfall regime. Unseen to most people, this change in rainfall affects the discharge of fine powders that used to decorate the new snow falling every year. Its compaction into coarse hail first and into ice later no longer occurs regularly, thus the particulate layers are disrupted that used to be geological markers, from which it was possible to date the glacier accretion just as for sediment layers (varves) deposited by water in a periglacial lake. The fine ash and aerosol layer from the 1815 Tambora eruption, encapsulated in the ice of glaciers all over the Earth, is a marker that contributed strongly to the quantitative evaluation of the climate change over three centuries, i.e., those before and after the beginning of the industrial revolution, besides giving clues on the movement of the prevailing winds. With the disappearance of the ice caps because of extensive melting, this important information for environmental studies would become unobtainable, and so equally impossible would be the extrapolation of results to study their deep past, as was done with some ice layers in Antarctica that turned out to have been deposited as long ago as nearly one million years.

To properly evaluate the fate of ice on an Earth that is undergoing climate change requires knowledge of a variety of local cases covering all possible environmental variables. In this Special Issue, the localities chosen for study are in Antarctica (Talos Dome), where the ice cap is very well preserved and could be drilled down to a depth that makes the assumption reasonable to have reached the layer from one million years ago (Baccolo et al., Maggi et al.); at the Ny-Ålesund station, Svalbard, as the representative location of the northern North Atlantic sector of the Arctic, where ice is so rapidly melting away as to open the sea route around the North Pole to trade ships (Ding et al.); and finally, in China (Liu et al., Xu et al.), with a particular emphasis on the entire Laohugou No.12 glacier system (glacier, soil, and moraine), i.e., in the typical situation of a high-mountain glacier that flows down to flat land inside a continent and melts slowly when reaching low enough altitudes.

In addition to different locations, this Special Issue considers it useful to clarify the ice problem by carefully describing different study methods, with a clear preference for the synchrotron-based ones since they allow measurements on very small samples, in turn containing exceedingly small amounts of particulate matter having chemically homogeneous properties (Cibin et al.). Indeed, among the novelty dealt with in this issue is the description of microdrop technology (Macis et al.), which involves melting ice into water and gives considerably better results than the usual filtering technique, because the evaporation of small droplets controls the distribution of the insoluble materials dispersed in the solution on a well-defined area according to a specific spatial pattern. This is particularly important in that it not only allows the application of a variety of spectroscopic methods, the radiation of which is not necessarily extracted from the synchrotron, but also gives clues on the possible derivation of their pollutants. On the other hand, the post-depositional biodegradation processes undergone by pollutant particles attacked by microbes while still lying on the ice surface and accumulating in characteristic holes within it (cryoconite) is a factor that indicates the complexity of the reactions at the very beginning of glacier formation (Pittino et al.).

Most particulate matter has been found to be iron-based, an element that is inferred to limit the biomass of phytoplankton populations in extensive regions of the ocean (i.e., the high-nutrient, low-chlorophyll regions, which are vital for feeding several marine species). In particular, minerals such as hematite (Fe_2O_3), magnetite (Fe_3O_4), and biotite, as well as the chemical compound ferrous oxalate dihydrate (FOD), have been detected. Their distribution is not casual, but depends on the land wherefrom they are surmised to originate (usually Antarctica, the Chilean Andes, and Tierra de Fuego, but perhaps southern Africa and Australia too) and on the depth of sampling of the specimen in the drilled ice core. Maggi et al. show that in their Antarctic source (TALDICE), the aeolian dust Fe-oxidation levels are higher than expected, probably because of their very high exposure levels, whereas they usually decrease in oxidation with the decreasing latitude of the inferred land of origin. By contrast, Baccolo et al. find that the upper levels of the Talos Dome ice core are essentially unaffected by depth, whereas the lowest ones (suggested to have their ice deposited before the Marine Isotope Stage-2 (MIS-2)) display incoherent variations due to the clustering of atoms, i.e., depending on their longer residence in the sequence.

Another major goal pursued in this issue is showing how beneficial to the understanding of climate change would be the results of a more intense cooperation between the synchrotron radiation community and the scientists working in polar areas. This goal is certainly reached—not only is the potential of upgraded laboratory equipment reviewed (Cappuccio et al.), but details and upgraded techniques in existing synchrotron apparatuses are also described (Cibin et al.), and new synchrotron lines are even described (e.g., LISA, by Puri et al., and the Chinese ones by Xu et al.). There is also a valuable example given of the simultaneous evaluation of several atom contents (Al, Ti, Si, Ca, Mg, Fe, Sr) determined in PM10, PM2.5, and PM1 at three characteristic environmental sites (Speranza et al.).

In conclusion, these three case studies may appear to be statistically insignificant, but they are only the start of a more extensive investigation, with each of them involving a very complex organization of work. The results on display show that the way is open to a better and increasingly deeper understanding of the climate change affecting the whole Earth—a phenomenon that humans must understand and quickly solve for their own survival.

Annibale Mottana

Accademia Nazionale dei Lincei and Accademia Nazionale delle Scienze detta dei XL

Rome, Italy

Editorial

Condensed Matter Researches in Cryospheric Science

Valter Maggi [1,2,*], **Cunde Xiao** [3,4] **and Augusto Marcelli** [5,6]

1   Earth and Environmental Sciences Department, University of Milano Bicocca, Piazza della Scienza, 1, I-20126 Milano, Italy
2   Istituto Nazionale di Fisica Nucleare, Sezione di Milano-Bicocca, Piazza della Scienza, 2, 20126 Milano, Italy
3   State Key Laboratory of Earth Surface Processes and Resource Ecology, Beijing Normal University, 19 Xinjiekouwai Street, Beijing 100875, China
4   State Key Laboratory of Cryospheric Sciences, Northwest Institute of Eco-Environment and Resources, Chinese Academy of Sciences, Lanzhou 730000, China
5   Istituto Nazionale di Fisica Nucleare, Laboratori Nazionali di Frascati, Via E. Fermi 40, I-00044 Frascati Rome, Italy
6   Rome International Centre for Material Science Superstripes, RICMASS, via dei Sabelli 119A, 00185 Rome, Italy
*   Correspondence: valter.maggi@unimib.it

Received: 5 July 2019; Accepted: 9 July 2019; Published: 12 July 2019

Keywords: cryospheric sciences; mineral dust; synchrotron light

1. Introduction

The comprehensive understanding of the cryosphere's global biogeochemical cycles represents a great challenge for the present climatic and environmental research on Earth. Many countries are involved in these challenging studies in different strategic areas at the Earth's poles and high mountain regions. China and Italy are strongly involved in these studies, with important results already obtained by their teams. The aim of this special issue, organized together by Chinese and Italian experts in the field, will cover climatic and environmental research studies, based on the detection and characterization of minerals and dust present in ice cores and aerosols in the atmosphere. This interdisciplinary, modern, and strategic research field looks at climate and pollution both at local and global scales [1,2].

Despite the increasing interest and great efforts, in particular over the last decade, there is a lack of consensus on many issues associated with environmental and climatic problems. The amount of studies regarding the environment, mountains and polar glaciers, and the cryosphere in general, are continuously increasing and yet remain far from reaching a conclusion. Ice cores, permafrost, and snow represent extraordinary climatic and environmental information archives that are seriously at risk because of the increasing temperatures on Earth. Research studies using new experimental methods may help in investigating the unique and precious archives with time and spatial resolutions, which were not even imaginable a few years ago. However, new ideas, methods, and approaches are required to improve and extend the characterization of ice and snow, which are extremely complex and fragile materials, and to investigate the very minor amounts of organic and inorganic materials hidden within them. These modern techniques could also be applied to other environmental problems, where the accurate detection and characterization of dust and aerosols present in the atmosphere are highly required.

A substantial improvement in the study of the physical and radiative characteristics of dust particles and in climate models has occurred in order to reliably predict their impact on climate. However, key properties of mineral particles are remain minimally known [3], in that global dust cycle simulations remain poor [4]. Where as, glacial archives, such as polar ice sheets or high-altitude mountain glaciers, may offer unique information on modern and past aeolian dust trapped within ice.

The background continental dust trapped in high-altitude glaciers and in polar ice sheets provides important information on dust transport in the troposphere, on the relative role of different source areas, and on wind strength. Specific events, such as volcanic eruptions or the impact of extraterrestrial materials [5,6], can be detected in ice cores, along with their effects on climate. These events can also be used to establish ice core chronologies [7].

Dust size distribution is a key issue together with particle morphology, mineralogical composition, and optical characteristics. Mineralogy of dust is required for source identification [8]. In addition, a large assemblage of mineral species within the same matrix and mixing with other poor crystalline fractions, such as volcanic products (i.e., tephra and ashes), organic particles (i.e., spores and bacteria), biomass burning products (i.e., soot), human emissions products (i.e., black carbon, sulfates), and extraterrestrial materials (i.e., micrometeorites) occur. Rare earth elements [9,10] and major elements [11] are related to dust provenance and environmental conditions at dust source areas. Due to the complex composition of dust, it is then mandatory for future climate research based on dust collected from deep ice cores to bring the ultimate limits of all available analytical methods to these studies.

2. Condensed Matter Applications in Cryospheric Sciences

2.1. Advanced Measurement Methods for Cryospheric Sciences

Due to the complexity of the interaction among cryosphere–atmosphere–lithosphere components, there are currently large uncertainties in the assessment of their physical–chemical properties and source apportionment. In addition, issues about their role in affecting glaciers/snow surfaces radiative effects, Earth's radiation budget, and environmental effects are still under investigation. With the aim of filling this gap, research on this topic has grown a lot in recent years and advanced experimental and modeling approaches have been proposed.

Large facilities, such as synchrotron radiation sources, offer new opportunities to investigations into cryospheric sciences, especially for the analysis of very low concentrations, as for the polar snow and glacier samples. For example, LISA (Linea Italiana per la Spettroscopia di Assorbimento di raggi X) is the beamline of the Italian Collaborating Research Group (CRG) at the European Synchrotron Radiation Facility (ESRF) dedicated to X-ray absorption spectroscopy (XAS). The beamline covers a wide energy range, $4 < E < 90$ keV, which offers the possibility to probe K and L edges of elements that are heavier than Ca [12].

Furthermore, Cibin et al. [13] developed instrumentation and protocols to optimize the collection of synchrotron radiation X-ray fluorescence, X-ray absorption spectroscopy, and X-ray powder diffraction data on ice core and possible source areas mineral dust samples, at the Diamond Light Source Facility (UK).

Many spectroscopic methods allow the characterization of the structure and electronic structure of samples, while the scattering/diffraction methods enable the determination of crystalline structures of either organic or inorganic systems. Moreover, imaging methods offer an unprecedented spatial resolution of samples, revealing their morphology and even their inner structure. In this issue, Xu et al. [14] introduce the synchrotron radiation facilities now available in mainland China, and the perspectives of synchrotron radiation-based methods suitable for investigating ice, snow, aerosols, dust, and other samples of cryospheric origin, i.e., deep ice cores, permafrost, filters, etc. The goal is to deepen the understanding of cryospheric sciences through increased collaboration between the synchrotron radiation community and scientists working in polar areas or involved in correlated environmental problems.

Nowadays, important research studies can be performed using conventional sources too. Indeed, the combination of low-power conventional sources and *polycapillary* optics allows the assembly of a prototype, which can provide a quasi-parallel intense beam for detailed X-ray spectroscopic analysis of extremely low concentrated samples. Cappuccio et al. [15] report the applications of the total external X-ray fluorescence (TXRF) station, a prototype assembled at the XLab Frascati laboratory (XlabF) at the

INFN National Laboratories of Frascati (INFN LNF). This laboratory has been established as a facility to study, design, and develop X-ray optics, in particular, polycapillary lenses, as well as to perform X-ray experiments for both elemental analysis and tomography.

The analysis of particulate matter (PM) in dilute solutions is an important target for environmental, geochemical, and biochemical research. Macis et al. [16] show how the microdrop technology may allow the control, through the evaporation of small droplets, of the deposition of insoluble materials dispersed in a solution on a well-defined area with a specific spatial pattern. Using this approach, the superficial density of the deposited solute can be accurately controlled, and it is possible to deposit an extremely reduced amount of insoluble materials, in the order of few µg on a confined area, thus allowing a relatively high superficial density to be reached within a limited time.

2.2. High Latitude-High Altitude Mineral Dust Atmospheric Transport

Mineral dust has a large impact on the Earth's radiative budget, ocean, and continental fertilization, as well as influencing many elemental biogeochemical cycles. Iron is well known to be a limit of the phytoplankton population's biomass in extensive regions of the ocean, which are referred to as high-nutrient low-chlorophyll (HNLC) regions, but iron speciation in continental soils is still poorly understood.

Liu et al. [17] investigated inorganic and organic standard substances, diluted mixtures of common Fe minerals in insoluble dust in snow from the Laohugou No.12 glacier, and sand including soil and moraine samples that were collected from western China using X-ray absorption near-edge structure (XANES) spectroscopy. Reference compounds showed that these samples contain only three mineral species: Fe_2O_3 (hematite), Fe_3O_4 (magnetite), biotite, and the chemical compound ferrous oxalate dihydrate (FOD), not yet recognized as a mineral species. These substances show a significant altitude effect depending on the elevation of the sampled snow samples.

Two contributions present the mineral dust characterization from Antarctic ice cores. Based on the XANES measurements, Fe K-edge spectra were collected on aeolian dust in the TALos Dome Ice CorE drilling project (TALDICE) ice core drilled in the peripheral East Antarctic plateau, as well as on Southern Hemisphere potential source area samples, by Maggi et al. [18]. While South American sources show, as expected, a progressive increase in Fe oxidation with decreasing latitude, Antarctic sources show Fe oxidation levels higher than expected in such a cold polar environment, probably because of their very high exposure ages. Results from the TALDICE dust samples are compatible with a South American influence at the site during the marine isotopic stage 2 (MIS2), the last and coldest phase of the last glacial period, in particular from Patagonia and Tierra del Fuego. However, a contribution from Australia and/or local Antarctic sources cannot be ruled out. Finally, important changes also occurred during the deglaciation and in the Holocene, when the influence of Antarctic local sources seems to become progressively more important in more recent times. Baccolo et al. [19] investigate the possibility of finding a stratigraphically intact ice sequence with a potential basal age exceeding one million years in Antarctica, and present here preliminary results on two sets of samples retrieved from the TALDICE ice core. A first set is composed of samples from the stratigraphically intact upper part of the core, the second by samples retrieved from the deeper part of the core that remains undated. Two techniques based on synchrotron radiation allowed characterization of the dust samples, showing that mineral particles entrapped in the deepest ice layers display altered elemental composition and anomalies concerning iron geochemistry, besides being affected by inter-particle aggregation.

Speranza et al. [20] applied compositional data analysis on mineral element concentrations, i.e., Al, Ti, Si, Ca, Mg, Fe, Sr content in PM_{10}, $PM_{2.5}$, and PM_1 simultaneous measurements at three characteristic environmental sites: kerbside, background, and rural site. Different possible sources of mineral trace elements affecting the PM in the considered sites were highlighted. Particularly, results show that compositional data analysis allows for the assessment of chemical/physical differences among mineral element concentrations of PM. These differences can be related to both different kinds of involved mineral sources and different mechanisms of accumulation/dispersion of PM at those sites.

2.3. Climatic Impact on the Cryospheric Environments

Ding et al. [21] report climate changes of Ny-Ålesund, Svalbard, a representative location of the northern North Atlantic sector of the Arctic, based on observational records from 1975 to 2014. Correlation among records of Ny-Ålesund and global HadCRUT4 datasets indicate the likelihood that the Arctic was experiencing a hiatus pattern, which just appeared later than at low to mid latitudes due to transport processes of atmospheric circulations and ocean currents, heat storage effect of cryospheric components, multidecadal variability of Arctic cyclone activities, etc. This case study provides a real new perspective on the global warming hiatus/slowdown debate.

Glaciers are important fresh-water reservoirs for our planet. Although they are often located at high elevations or in remote areas, glacial ecosystems are not pristine, as many pollutants can undergo long-range atmospheric transport and be deposited on glacier surface, where they can be stored for long periods of time, and then be released into the down-valley ecosystems. Pittino et al. [22] review studies on cryoconite holes, which occur on the surface of most glaciers. They are small ponds filled with water and a layer of sediment, named cryoconite, at the bottom. Indeed, these are hotspot environments for biodiversity on glacier surface as they host metabolically active bacterial communities that include generalist taxa able to degrade pollutants. These studies have also revealed that bacteria play a significant role in pollutant degradation in these habitats and can be positively selected in contaminated environments.

3. Concluding Remarks

These results present highlights of some of the most recent advances in cryospheric studies, especially in relation to mineral dust and aerosols in the atmosphere. They evidence the complexity of chemical–physical processes involving solid compounds occurring in glacier, snow, and permafrost environments, covering different aspects such as spatial and temporal trends, as well as the impact of the mineral and non-mineral particles. These studies also demonstrate the need for collaborative interdisciplinary and transnational efforts to better understand the challenges of the present climatic and environmental research studies on Earth, but also out of the Earth's system. The results show that recent advances in measurement techniques and source apportionment are powerful and sophisticated tools that may provide novel high-quality scientific information but represent only the first challenging step.

Author Contributions: All the authors contributed equally.

Funding: This research received no external funding.

Acknowledgments: We express our thanks to all authors that contributed to this special issue, to the journal *Condensed Matter* that hosts these contributions and to the MDPI staff for their continuous support.

Conflicts of Interest: The authors declare no conflict of interest.

References

1. Schulz, M.; Prospero, J.M.; Baker, A.R.; Dentener, F.; Ickes, L.; Liss, P.S.; Mahowald, N.M.; Nickovic, S.; García-Pando, C.P.; Rodríguez, S.; et al. Atmospheric Transport and Deposition of Mineral Dust to the Ocean: Implications for Research Needs. *Environ. Sci. Technol.* **2012**, *46*, 10390–10404. [CrossRef] [PubMed]

2. Shi, Z.B.; Krom, M.D.; Jickells, T.D.; Bonneville, S.; Carslaw, K.S.; Mihalopoulos, N.; Baker, A.R.; Benning, L.G. Impacts on iron solubility in the mineral dust by processes in the source region and the atmosphere: A review. *Aeolian Res.* **2012**, *5*, 21–42. [CrossRef]

3. Formenti, P.; Schutz, L.; Balkanski, Y.; Ebert, M.; Kandler, K.; Petzold, A.; Scheuvens, D.; Weinbruch, S.; Zhang, D. Recent progress in understanding physical and chemical properties of African and Asian mineral dust. *Atmos. Chem. Phys.* **2011**, *11*, 8231. [CrossRef]

4. Huneeus, N.; Schulz, M.; Balkanski, Y.; Griesfeller, J.; Prospero, J.; Kinne, S.; Bauer, S.; Boucher, O.; Chin, M.; Dentener, F.; et al. Global dust model intercomparison in AeroCom phase I. *Atmos. Chem. Phys.* **2011**, *11*, 7781. [CrossRef]

5. Narcisi, B.; Petit, J.R.; Engrand, C. First discovery of meteoritic events in deep Antarctic (EPICA-Dome C) ice cores. *Geophys. Res. Lett.* **2007**, *34*, L15502. [CrossRef]

6. Narcisi, B.; Petit, J.R.; Delmonte, B.; Basile-Doelsch, I.; Maggi, V. Characteristics and sources of tephra layers in the EPICA-Dome C ice record (East Antarctica): Implications for past atmospheric circulation and ice core stratigraphic correlations. *Earth Planet. Sci. Lett.* **2005**, *239*, 253. [CrossRef]
7. Parrenin, F.; Barnola, J.M.; Beer, J.; Blunier, T.; Castellano, E.; Chappellaz, J.; Dreyfus, G.; Fischer, H.; Fujita, S.; Jouzel, J.; et al. The EDC3 chronology for the EPICA Dome, C. ice core. *Clim. Past* **2007**, *3*, 485. [CrossRef]
8. Maggi, V. Mineralogy of atmospheric microparticles deposited along the Greenland Ice Core Project ice core. *J. Geophys. Res.* **1997**, *102*, 725. [CrossRef]
9. Wegner, A.; Gabrielli, P.; Wilhelms-Dick, D.; Ruth, U.; Kriews, M.; De Deckker, P.; Barbante, C.; Cozzi, G.; Delmonte, B.; Fischer, H. Change in dust variability in the Atlantic sector of Antarctica at the end of the last deglaciation. *Clim. Past* **2012**, *8*, 135–147. [CrossRef]
10. Gabrielli, P.; Wegner, A.; Petit, J.R.; Delmonte, B.; De Deckker, P.; Gaspari, V.; Fischer, H.; Ruth, U.; Kriews, M.; Boutron, C.; et al. A major glacial-interglacial change in aeolian dust composition inferred from Rare Earth Elements in Antarctic ice. *Quat. Sci. Rev.* **2010**, *29*, 265. [CrossRef]
11. Marino, F.; Castellano, E.; Ceccato, D.; De Deckker, P.; Delmonte, B.; Ghermandi, G.; Maggi, V.; Petit, J.R.; Revel, M.; Udisti, R. Defining the geochemical composition of the EPICA Dome C ice core dust during the last glacial-interglacial cycle. *Geochem. Geophys. Geosyst.* **2008**, *9*, Q10018. [CrossRef]
12. Puri, A.; Lepore, G.O.; d'Acapito, F. The New Beamline LISA at ESRF: Performances and Perspectives for Earth and Environmental Sciences. *Condens. Matter* **2019**, *4*, 12. [CrossRef]
13. Cibin, G.; Marcelli, A.; Maggi, V.; Baccolo, G.; Hampai, D.; Robbins, P.; Liedl, A.; Polese, C.; D'Elia, A.; Macis, S.; et al. Synchrotron Radiation Research and Analysis of the Particulate Matter in Deep Ice Cores: An Overview of the Technical Challenges. *Condens. Matter* **2019**, *4*, 61. [CrossRef]
14. Xu, W.; Du, Z.; Liu, S.; Zhu, Y.; Xiao, C.; Marcelli, A. Perspectives of XRF and XANES Applications in Cryospheric Sciences Using Chinese SR Facilities. *Condens. Matter* **2018**, *3*, 29. [CrossRef]
15. Cappuccio, G.; Cibin, G.; Dabagov, S.B.; Di Filippo, A.; Piovesan, G.; Hampai, D.; Maggi, V.; Marcelli, A. Challenging X-ray Fluorescence Applications for Environmental Studies at XLab Frascati. *Condens. Matter* **2018**, *3*, 33. [CrossRef]
16. Macis, S.; Cibin, G.; Maggi, V.; Baccolo, G.; Hampai, D.; Delmonte, B.; D'Elia, A.; Marcelli, A. Microdrop Deposition Technique: Preparation and Characterization of Diluted Suspended Particulate Samples. *Condens. Matter* **2018**, *3*, 21. [CrossRef]
17. Liu, S.; Xiao, C.; Du, Z.; Marcelli, A.; Cibin, G.; Baccolo, G.; Zhu, Y.; Puri, A.; Maggi, V.; Xu, W. Iron Speciation in Insoluble Dust from High-Latitude Snow: An X-ray Absorption Spectroscopy Study. *Condens. Matter* **2018**, *3*, 47. [CrossRef]
18. Maggi, V.; Baccolo, G.; Cibin, G.; Delmonte, B.; Hampai, D.; Marcelli, A. XANES Iron Geochemistry in the Mineral Dust of the Talos Dome Ice Core (Antarctica) and the Southern Hemisphere Potential Source Areas. *Condens. Matter* **2018**, *3*, 45. [CrossRef]
19. Baccolo, G.; Cibin, G.; Delmonte, B.; Hampai, D.; Marcelli, A.; Di Stefano, E.; Macis, S.; Maggi, V. The Contribution of Synchrotron Light for the Characterization of Atmospheric Mineral Dust in Deep Ice Cores: Preliminary Results from the Talos Dome Ice Core (East Antarctica). *Condens. Matter* **2018**, *3*, 25. [CrossRef]
20. Speranza, A.; Caggiano, R.; Pavese, G.; Summa, V. The Study of Characteristic Environmental Sites Affected by Diverse Sources of Mineral Matter Using Compositional Data Analysis. *Condens. Matter* **2018**, *3*, 16. [CrossRef]
21. Ding, M.; Wang, S.; Sun, W. Decadal Climate Change in Ny-Ålesund, Svalbard, A Representative Area of the Arctic. *Condens. Matter* **2018**, *3*, 12. [CrossRef]
22. Pittino, F.; Ambrosini, R.; Azzoni, R.S.; Diolaiuti, G.A.; Villa, S.; Gandolfi, I.; Franzetti, A. Post-Depositional Biodegradation Processes of Pollutants on Glacier Surfaces. *Condens. Matter* **2018**, *3*, 24. [CrossRef]

Article

Decadal Climate Change in Ny-Ålesund, Svalbard, A Representative Area of the Arctic

Minghu Ding [1,2], Shujie Wang [1,3] and Weijun Sun [3,*]

[1]   Institute of Polar Meteorology, Chinese Academy of Meteorological Sciences, Beijing 100081, China; dingmh@cma.gov.cn (M.D.); wangshujie1990@foxmail.com (S.W.)
[2]   State Key Laboratory of Cryospheric Sciences, Cold and Arid Regions Environmental and Engineering Research Institute, Chinese Academy of Sciences, Lanzhou 730000, China
[3]   College of Geography and Environment, Shandong Normal University, Jinan 250014, China
*   Correspondence: 612033@sdnu.edu.cn

Received: 23 February 2018; Accepted: 3 April 2018; Published: 8 April 2018

Abstract: In recent decades, global warming hiatus/slowdown has attracted considerable attention and has been strongly debated. Many studies suggested that the Arctic is undergoing rapid warming and significantly contributes to a continual global warming trend rather than a hiatus. In this study, we evaluated the climate changes of Ny-Ålesund, Svalbard, a representative location of the northern North Atlantic sector of the Arctic, based on observational records from 1975–2014. The results showed that the annual warming rate was four times higher than the global mean (+0.76 °C·decade^{-1}) and was also much greater than Arctic average. Additionally, the warming trend of Ny-Ålesund started to slow down since 2005–2006, and our estimates showed that there is a 8–9 years-lagged, but significant, correlation between records of Ny-Ålesund and global HadCRUT4 datasets. This finding indicates that the Arctic was likely experiencing a hiatus pattern, which just appeared later than the low-mid latitudes due to transport processes of atmospheric circulations and ocean currents, heat storage effect of cryospheric components, multidecadal variability of Arctic cyclone activities, etc. This case study provides a new perspective on the global warming hiatus/slowdown debate.

Keywords: Arctic; Arctic rapid warming; global warming hiatus; global warming slowdown

---

1. Introduction

The Arctic, Antarctica, and the Tibetan Plateau are the most sensitive areas to climate change and are, therefore, considered as key components in global change estimates, including those of the WCRP (World Climate Research Program) and IGBP (International Geosphere-Biosphere Program). These areas are also of major concern in all IPCC (Intergovernmental Panel on Climate Change) reports. Compared with the attention given to Antarctica and the Tibetan Plateau, greater attention has been paid to the rapid transitions in the Arctic over the past few decades and their complex ecological and economic impacts on surrounding countries. Additionally, several international projects, including PPP (http://www.polarprediction.net/) and MOSAiC (http://www.mosaicobservatory.org/), have been launched to gain a better understanding of the Arctic.

The surface air temperature in the Arctic has experienced warming at a rate of more than twice the global average, a phenomenon known as Arctic amplification [1–3]. However, Arctic warming has not occurred linearly; it primarily occurred from 1920–1940 and continued from 1970 to the present, whereas pronounced cooling occurred between the two warming periods [4–6]. Many attempts have been made to interpret Arctic amplification based on the surface albedo feedback related to the decrease in snow depth, retreating sea ice extent [7,8], increased meridional heat advection to the high latitudes resulting from atmospheric and oceanic circulations [6,9–11], and systematic changes in water vapor and cloud cover above the Arctic [12]. However, even when incorporating these processes into climate models, climate changes in polar areas remain difficult to predict, which indicates that the underlying mechanisms of polar amplification remain uncertain and debatable [13,14].

In recent years, a large number of rapid climate changes in the Arctic have greatly affected a broad spectrum of physical, ecological, and economic systems [1,15]. The volume of Arctic sea ice has decreased by 75% since the 1980s [16,17], and the Artic will likely be nearly seasonally ice-free before 2050 [17]. The rapid reduction of Arctic sea ice plays a crucial role in the regional climate over the Arctic and even the climate and weather at lower latitudes [13]. For example, autumn-winter Arctic sea ice anomalies in 2012 affected the development of the Siberian High and resulted in extreme weather events in the winter over Eurasia [18,19]. Ten years ago, it was reported that climate change of the Arctic could be aggravated and further constrain the evolution of the global climate system in this century [20], but no authors predicted such a rapid transition as that occurring currently.

In comparison with Antarctica, the land surface of the Arctic is more complex and includes ice sheets/caps, glaciers, and frozen ground. Furthermore, climatic and environmental changes in these areas vary largely over local and regional scales and play crucial roles in feedback mechanisms [21]. Therefore, it is highly relevant to evaluate climatic changes in these regions and their similarities and differences with global average conditions. To better understand these changes, long-term and high-quality meteorological observations are key.

In this paper, climate change in Ny-Ålesund, is evaluated using meteorological records from three stations (the Yellow River Station (YRS), Norwegian Polar Institute (NPI) and the joint French-German Arctic Research Base of the Alfred Wegener Institute for Polar and Marine Research and Polar Institute Paul Emile Victor (AWIPEV)). A further analysis and discussion are also presented to assess the related contributing factors.

2. Study Area

Ny-Ålesund (78°55′ N, 11°56′ E) is located at the western coast of Svalbard, which is one of the northernmost archipelagos in the Arctic (Figure 1). The Svalbard archipelago consists of Barentsøya, Nordaustlandet, and Spitsbergen, covering a total area of 62,700 km^2, and glaciers cover 60% of the surface area. The James Island ice sheet lies to the east of Ny-Ålesund. This area is characterized by a humid climate resulting from the North Atlantic warming current, and winter daily mean temperatures can exceed 0 °C sometime [22]. Although Ny-Ålesund has distinct climate conditions, it can also provide evidence for climate change throughout the Arctic, especially the Atlantic sector of the Arctic, as demonstrated in the present study.

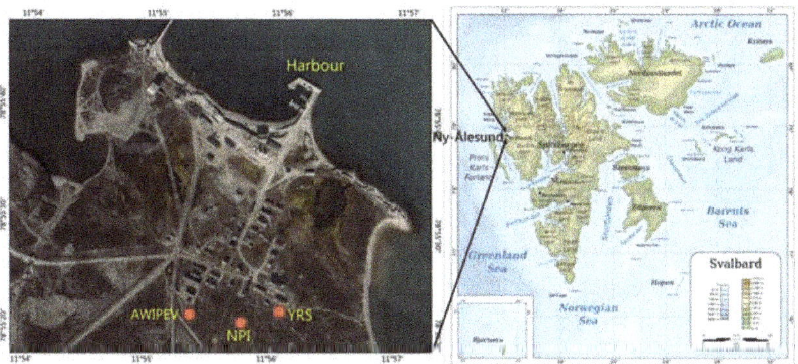

Figure 1. Aerial photo and map of Ny-Ålesund, Svalbard, Arctic.

3. Data Description

The data used in this paper are derived from three automatic weather stations located at Ny-Ålesund, Svalbard (Figure 1). The Sverdrup Research Station was established by the NPI, and has daily meteorological observations from 1975 to 2014 (data available on http://sharki.oslo.dnmi.no). The AWIPEV operates a 10 m meteorological tower locating on a field of soft tundra and provides 5-min resolution observational data from 1994 to 1998 and 1-min resolution data from 1998 to 2011 (data available on the PANGAEA repository, doi:10.1594/PANGAEA.793046). China established the YRS in 2003, and has an automated weather station located in the eastern portion of the township near the bird sanctuary. This station can provide air temperature, relative humidity, air pressure, wind speed and wind direction data, at a 1-h resolution and a height of 2 m from 2005 to 2014 (Table 1, data available on http://www.chinare.org.cn/index/). The horizontal distances from the YRS to the AWIPEV and NPI stations are 750 m and 400 m, respectively. These three stations collect observations according to WMO (World Meteorological Organization) standards with reliable sensors (Table 1 for example). To guarantee the continuity of the observational data, we used 2-m height observational data from the three meteorological stations. Note that relative humidity and wind speed data at 2 m were missing from January 2005 to August 2005, and wind speed data at 2 m of YRS were missing during 2014.

Table 1. Specifications of the automated weather Yellow River Station at Ny-Ålesund.

Element	Sensor Type	Measurement Range	Accuracy
Air temperature, °C	Vaisala HMP155	−80–60 °C	±0.2 °C
Relative humidity, %	Vaisala HMP155	0–100%	±1.7%
Air pressure, hPa	Vaisala PTB220	500–1100 hPa	±0.3 hPa
Wind speed, $m \cdot s^{-1}$	XFY3-1	$0–90\ m \cdot s^{-1}$	$±0.5\ m \cdot s^{-1}$
Wind direction, °	XFY3-1	0–360°	±5°

Although the three meteorological data sets do not exactly coincide due to the different altitudes, the records of the three meteorological stations were consistent with each other, including the air temperature records, and the correlation coefficients exceeded 0.98 ($p < 0.01$) among the hourly records of NPI, AWIPEV, and YRS from 2006–2010.

Control criteria similar to those proposed by Maturilli et al. [22] were applied. If more than 30 min of data were missing within 1 h, we considered the hourly mean value to be a missing data point. If more than 5 h of data were missing over one day, we considered the daily mean value to be a missing data point. If more than four days of data were missing within one month, we considered the monthly mean value to be a missing data point. Considering these durations, there were no missing monthly mean values in our datasets. Local standard time (LST, 2 h earlier than GMT) was used in this study.

Compared to relative humidity, specific humidity better reflects annual variations in water vapor in the atmosphere and is calculated using the following equation:

$$q = \frac{622e}{p - 0.378e},$$

(1)

where q is specific humidity, p is air pressure, and e is the water vapor pressure.

For convenience, we defined spring as March–May, summer as June–August, autumn as September–November, and winter as December–February.

4. Results

4.1. The Decadal Climate Change of Ny-Ålesund

Based on the records used in this study, we found that air temperatures were relatively low from 1977 to 1983 (Figure 2a), although 1988 was the coldest year (-8.5 °C). In addition, the overall annual mean temperatures reflected a significant warming of 0.76 ± 0.29 °C decade^{-1} from 1975 to 2014 ($p < 0.05$). Furthermore, the air temperature increased in all seasons, with the largest increases occurring in winter and spring. The warming rates in spring, summer, autumn and winter were 0.58 ± 0.5 °C·decade^{-1}, 0.27 ± 0.01 °C·decade^{-1}, 0.49 ± 0.04 °C·decade^{-1}, and 1.17 ± 0.07 °C·decade, respectively (Figure 3). This pattern is similar to the proxy trend at Svalbard Airport reconstructed by [23,24] reconstructed the air temperature in the northern Atlantic-Arctic region (60° W–45° E, 60°–90° N) from 1802 to 2009 using CRUTEM3v data and four long-term observational temperature series as predictors (from Tornedalen, Iceland, Southwest Greenland, and Archangelsk). Their results exhibited a trend similar to that in Ny-Ålesund, but their estimated warming rate was much lower.

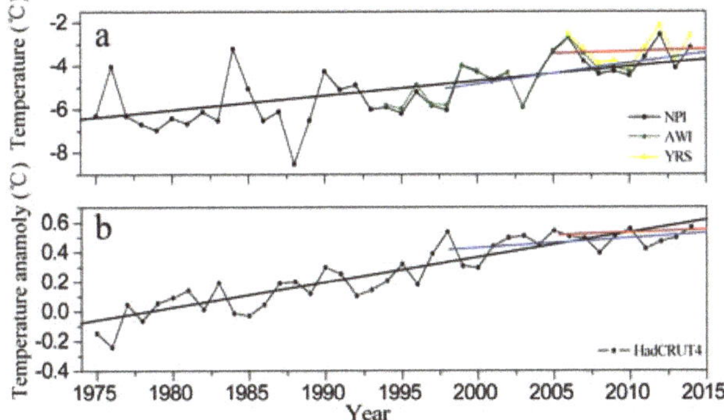

Figure 2. Comparison of the annual mean surface air temperatures at Ny-Ålesund (**a**) with the HadCRUT4 (**b**) global record. The straight black lines, blue lines, and red lines are the linearly-regressed trends.

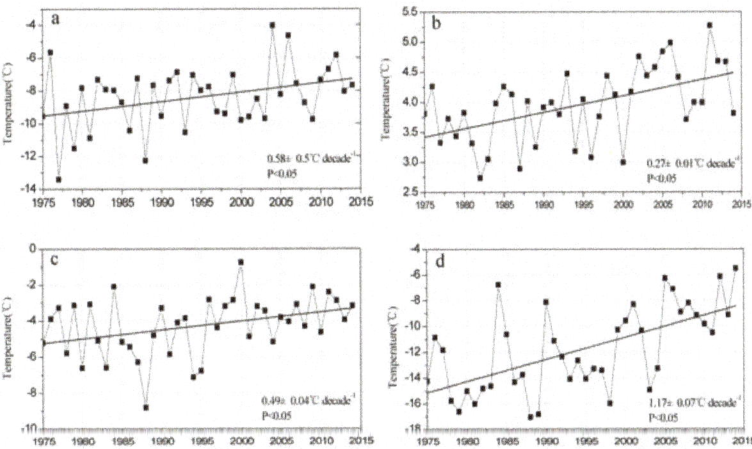

Figure 3. Inter-annual variations and linear trends in air temperature for winter (**a**), spring (**b**), summer (**c**), and autumn (**d**) at Ny-Ålesund from 1975 to 2014.

Compared with other areas such as Svalbard Airport, Hopen, or Bjørnøya [25], Ny-Ålesund is not the warmest area. Some coastal areas in Greenland also exhibited a rapid warming during past 20 years [26]. Especially the west coast, the air temperature increased ~10 °C during 1991–2012. However, there were also some areas that exhibited cooling trends, such as the southern part of Greenland.

In recent years, the global warming hiatus/slowdown has been debated. The global warming rates during the early 2000s were lower than previous rates [27], and many studies found that the main reason should be the absence or wrong assimilation of the dataset in polar areas [28]. To evaluate whether a warming slowdown occurred in Ny-Ålesund, we compared the temperature variations in Ny-Ålesund with global variations from 1975 to 2014. Figure 2 shows that the warming rate of the annual mean temperature in Ny-Ålesund was approximately four times greater than the global average (0.76 ± 0.29 °C·decade^{-1} vs. 0.17 ± 0.03 °C·decade^{-1}, $p < 0.05$). From 1998–2014, the warming rate in Ny-Ålesund was 1.04 ± 0.84 °C·decade^{-1} ($p < 0.05$), whereas the global mean was 0.06 ± 0.08 °C decade^{-1} ($p < 0.10$), which is agreed with the pattern studied by Huang et al. [28]. In the last decade (2005–2014), the warming rate in Ny-Ålesund slowed to 0.03 ± 1.85 °C·decade^{-1} but was still higher than the global average of 0.01 ± 0.15 °C decade^{-1} (both did not pass the 95% significance test). This trend suggests that warming in Ny-Ålesund has been far greater than that of the global mean temperature and also the Arctic mean; in other words, Arctic amplification is impactful on both long- and short-term scales. From Figure 2 we can also find a warming slowdown also occurred in Ny-Ålesund, but lagged behind the decrease in the global warming rate, which will be discussed in the next part.

Similar with air temperature, specific humidity also exhibited a positive trend of 0.14 ± 0.12 g·kg^{-1}·decade^{-1} (Figure 4a) during 1994–2014, but, in summer, it was relatively stable for the strong large-scale advection of large atmospheric circulation and local evaporation effects. The air pressure differed and exhibited considerable month-to-month variations (Figure 4b), and has no significant change from 1994–2014, same with wind speed (Figure 4c).

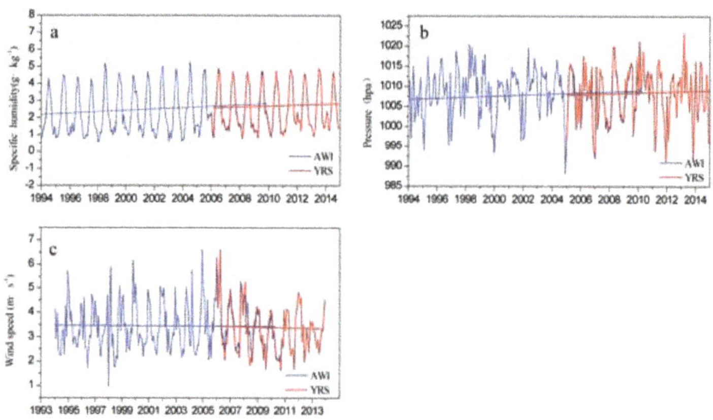

Figure 4. Variations of monthly mean specific humidity (**a**), air pressure (**b**); and wind speed (**c**); the blue and red straight lines represent the linear trends.

4.2. How Long does the Climate Change of Ny-Ålesund Lag Behind the Global Change?

Lead-lag analysis was conducted to identify the correlation between climate change of Ny-Ålesund and global mean (Figure 5). The result shows that the Ny-Ålesund and global temperature variations were remarkably consistent, with a lag time of 8–9 years. This implies that the "warming hiatus" many scientists studied also appears in Ny-Ålesund, it just started later than the other areas. Furthermore, it can be concluded that the Arctic can not only amplify global climate changes, but also regulates climate change by being an energy reservoir.

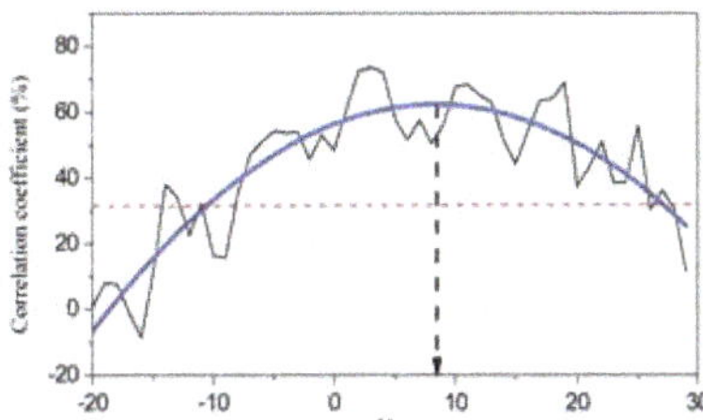

Figure 5. The lead-lag correlation analysis between the Ny-Ålesund and global air temperatures calculated using the annual mean air temperatures from NPI station and HadCRUT4 during 1975–2014. The black line represents the correlation coefficient, and the thick blue line represents a first-order polynomial fitting. The red dashed line indicates the 95% confidence level.

4.3. Can Ny-Ålesund Represent the Arctic?

As discussed in Section 4.1, the warming rates in Ny-Ålesund were calculated to be relatively high. However, the spatial representativeness of Ny-Ålesund is a key question to discuss the climate change of Arctic. To answer this question, ERA-interim reanalysis data was used for correlation analyses in the following context.

As shown in Figure 6, the correlation of air temperature, specific humidity and air pressure distributions between Ny-Ålesund and Arctic were calculated and exhibited large scale coherence. Especially air temperature, the record of Ny-Ålesund can capture the variation of surface temperature over most of Arctic. In one word, the observations of Ny-Ålesund can partly represent the northern North Atlantic sector of the Arctic.

In addition, ERA-interim was able to capture the inter-annual variability in wind speed at Ny-Ålesund ($r > 0.5$, $p < 0.05$), but there were no obvious and significantly correlated regions. This finding reveals that the wind of Ny-Ålesund is mainly influenced by local morphology.

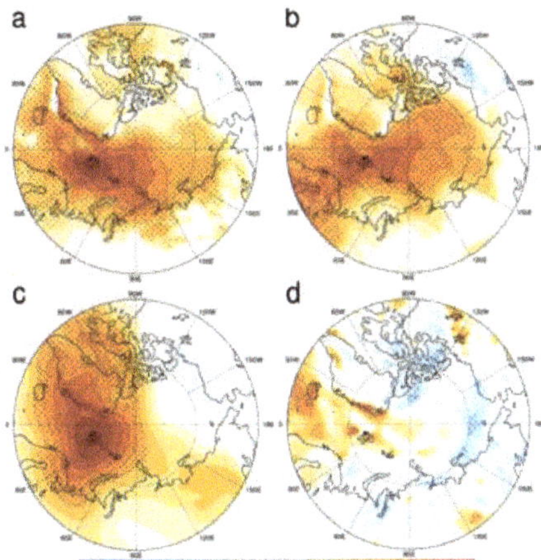

Figure 6. The spatial correlation patterns between the observed and ERA-interim modeled (**a**) 2 m temperatures, (**b**) specific humidity, (**c**) air pressures; and (**d**) 10 m winds with annual resolution during 1979–2014 for 2 m temperature, and 1994–2010 for specific humidity, air pressure, 10 m winds. The stippling indicates the 95% confidence level.

5. Discussion

Compared with global climate change and Arctic climate change, Ny-Ålesund can be defined as an ART (Arctic Rapid Transition) area for its rapid warming from 1975–2014. However, the variations of air temperature lagged behind global variations by 8–9 years. Consequently, two questions arise: why does the climate of Ny-Ålesund change so rapidly, and why does the climate change of Ny-Ålesund lag behind global change by 8–9 years?

5.1. Why is the Climate of Ny-Ålesund Changes so Rapid?

As many studies on Arctic amplification have suggested, one cause of rapid climate change in the region is sea ice retreat [13]. Since 1979, the winter sea ice area north of Svalbard has diminished by 10% decade^{-1} [29]. The ice-albedo feedback mechanisms associated with the sea ice extent and changes of surface heat fluxes are responsible for Arctic amplification [7,8,30]. However, similar processes have also occurred in other Arctic regions; thus, this reason should not be the most important one for the rapid change.

Atmospheric cyclones can transport energy from low to high latitudes. Previous studies [31,32] have analyzed the frequencies and air pressures of cyclones that entered or formed within the Arctic basin, and found that, since the 1950s, the number of cyclones entering the Arctic basin has increased significantly (but the frequency of Arctic cyclones that formed within the Arctic basin did not). Meanwhile, the frequency of deep cyclones entered and formed within the basin also increased. These systems allow more humidity and heat to be transported to the Arctic. Furthermore, the most significant changes of seasonal parameters associated with cyclones occurred in winter [32] and have led to stronger effects on winter climate, including storms.

Although the process is slow, ocean currents can deliver more energy to polar areas due to their large heat capacities. Evidence has suggested that the northward flow of warm Atlantic water has been enhanced in recent decades [33–36] and is the major cause of heat advection that strongly affects the sea ice distribution. From observations, the surface temperature of Atlantic water has increased by 1.1 °C [29]. Ny-Ålesund locates in the Atlantic Arctic region, and is surrounded by the West Spitsbergen Current which transports advective oceanic heat from low latitudes [37]. It may play one of the most important roles in shaping local climate conditions.

A recent study [38] noted that sea ice retreat in the Barents-Kara Sea area can affect the strength and position of the polar vortex and increase the frequency of blocking regimes over the Euro-Atlantic sector in late winter. Circulation regimes in the North Atlantic may also change, which can induce local warming.

Due to sea ice retreat and more open ocean surface, there has been pronounced increase of tropospheric water vapor above the Arctic, by studies based on radiosondes, reanalysis data, or satellite retrievals [39–42]. These estimations are consistent with records of specific humidity of Ny-Ålesund. It will accelerate the warming of Arctic due to the greenhouse gas effect, whereas the stratospheric water vapor has decreased and it may cool the air of Arctic [43,44]. Additionally, these large-scale backgrounds, the frozen polar tundra domain of the land surface of Svalbard, can absorb more energy than the ocean during the polar summer due to its higher heat capacity. This seasonal regulation effect can enhance the warming during winter, and is a reason for the high climate sensitivity of Ny-Ålesund.

5.2. Why Climate Change of Ny-Ålesund Lags Global Change by 8–9 Years?

As we all know, most solar radiation absorbed by the Earth occurs at low latitudes, and as a consequence the global warming starts at low latitudes. The energy heating polar areas is transported by atmospheric circulations, generated by a temperature gradient between the tropics and poles.

Under the influence of global climate warming, atmospheric patterns including intensities of Arctic cyclones and Arctic sea level pressures [4,32] has changed. This may also affect the variation of air temperature. Polyakov et al. [4] studied the relation between air temperature and cyclone activity in the Arctic, and found that the Arctic air temperature is characterized by low-frequency oscillations, especially strong multidecadal variability, and lags behind cyclone activity by 5–15 years. This is in agreement with our estimate.

Nevertheless, some studies found that the extent of the Siberian High has increased and moved northward since 2004, the associated meridional wind has been significantly enhanced and more warm air masses have been transported to the Arctic [45]. This will strengthen the warming of the Arctic after 2004. In addition to atmospheric circulation, large-scale sea circulation can also promote warm air and water in the Atlantic-Arctic region and influence the regional climate. Figure 7 shows the spatial patterns of correlation between annual mean air temperature of Ny-Ålesund and the North Atlantic sea surface temperature (SST), and it can be found that the two are well correlated. This demonstrates that air masses/energy above the North Atlantic can be easily transported by cyclones or storms. However, the ocean current transfer is much slower. Warm surface currents take a long time to travel to polar areas (their interactions with surface air may accelerate the process). Controlled by trade winds and decadal changes of deeper layers over the Atlantic and Southern Oceans, the SST of the equatorial region is characterized by internal decadal variation and affects the energy distribution over the Earth's surface [46–49], whereas there is no clear mechanism for the Arctic. However, the latest results showed that the SST anomaly of the northern Nordic Seas lags behind the North Atlantic by ~10 years [50,51], and they suggested that the SST advection of Norwegian Atlantic current should be one important factor. Their findings also provide evidence for the delayed warming slowdown in Ny-Ålesund.

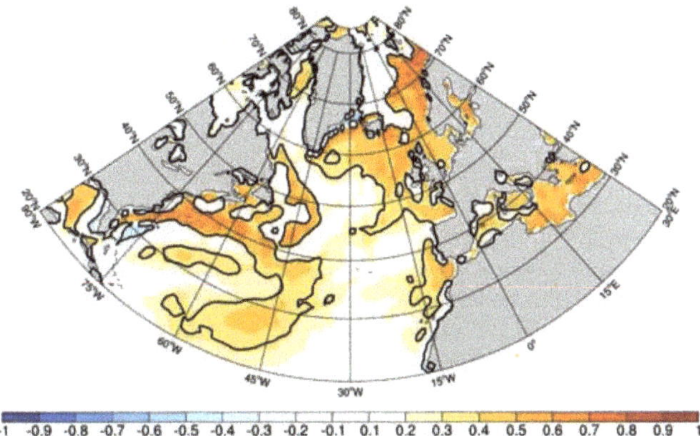

Figure 7. The spatial correlation pattern between the annual mean air temperature at Ny-Ålesund and the North Atlantic annual sea surface temperature (SST).

Sea ice, glaciers, ice sheet/caps, and frozen soil in the Arctic are sensitive components to climate change and have been retreating for years. These changes and high heat capacity can delay or slow variations in other components. In other words, energy reservoirs of these components are also key factors associated with the lagged changes in Arctic climate. However, quantifying the contributions of these factors on the decreased rate of warming is difficult, and further research is required to understand the detailed mechanisms.

6. Conclusions

In this paper, we estimate the variations of air temperature, specific humidity, air pressure, and wind speed during 1975–2014 in Ny-Ålesund, a typical area of frozen tundra in the Arctic. The linear trend of warming was $0.76 \pm 0.29\ °C \cdot decade^{-1}$, four times greater than the global mean. This rapid warming is a comprehensive result of atmospheric activity, retreating sea ice, regional ocean-air interactions, and surface conditions, as many studies pointed out.

Although the continuous warming since the mid-1990s is inconsistent with the "hiatus", we find that there was a warming hiatus/slowdown since 2005 at Ny-Ålesund. Additionally, the variation of air temperature lags by 8–9 years, which implies that the warming hiatus probably exists in the Arctic but lags behind, globally. This phenomenon is not an isolated instance, An et al. [52] reported that the warming rate above 4000 m of the Tibetan Plateau has been slowing since the mid-2000s. In the Antarctic Peninsula, the slowdown of the increasing temperature trend was also found after 1998/1999, however, the reason is attributed to local phenomena, such as the deepening of Amundsen Sea Low and not due to the global hiatus [53].

From the correlation analysis, we found Ny-Ålesund could represent most Arctic areas, especially the Atlantic-Arctic sector. Therefore, this warming hiatus may also appear in other Arctic areas and convert the estimation of global change. The oscillations of atmospheric dynamic systems, the methods of energy transport from low to high latitudes, and feedback mechanisms of the Artic on climate change may contributes to the warming hiatus. However, more evidence is required and further studies need to be carried out to make this clear.

Acknowledgments: The research received financial assistance from the National Natural Science Funding of China (41771064, 41690143) and the basic funding from CAMS.

Author Contributions: W.S. provided the idea and designed the study. S.W. and M.D. illustrated the figures. M.D. wrote the manuscript.

Conflicts of Interest: The authors declare no conflict of interest.

References

1. Jeffries, M.O.; Overland, J.E.; Perovich, D.K. The arctic shifts to a new normal. *Phys. Today* **2013**, *66*, 35. [CrossRef]
2. Serreze, M.C.; Francis, J.A. The arctic amplification debate. *Clim. Chang.* **2006**, *76*, 241. [CrossRef]
3. Johannessen, O.M.; Kuzmina, S.I.; Bobylev, L.P.; Miles, M.W. Surface air temperature variability and trends in the arctic: New amplification assessment and regionalisation. *Tellus* **2016**, *68*, 28234. [CrossRef]
4. Polyakov, I.V.; Bekryaev, R.V.; Alekseev, G.V.; Bhatt, U.S.; Colony, R.L.; Johnson, M.A. Variability and trends of air temperature and pressure in the maritime arctic, 1875–2000. *J. Clim.* **2003**, *16*, 2067. [CrossRef]
5. Johannessen, O.M.; Bengtsson, L.; Miles, M.W.; Kuzmina, S.I.; Semenov, V.A.; Alekseev, G.V. Arctic climate change—Observed and modeled temperature and sea ice. *Tellus* **2004**, *56*, 328. [CrossRef]
6. Chylek, P.; Folland, C.K.; Lesins, G.; Dubey, M.K.; Wang, M. Arctic air temperature change amplification and the atlantic multidecadal oscillation. *Geophys. Res. Lett.* **2009**, *36*, 61. [CrossRef]
7. Serreze, M.C.; Barrett, A.P.; Stroeve, J.C.; Kindig, D.N. The emergence of surface-based arctic amplification. *Cryosphere. Discuss.* **2009**, *3*, 11. [CrossRef]
8. Screen, J.A.; Simmonds, I. The central role of diminishing sea ice in recent arctic temperature amplification. *Nature* **2010**, *464*, 1334. [CrossRef] [PubMed]
9. Bengtsson, L.; Semenov, V.A.; Johannessen, O.M. The early twentieth-century warming in the arctic—A possible mechanism. *J. Clim.* **2004**, *17*, 4045. [CrossRef]
10. Hwang, Y.; Frierson, D.M.W.; Kay, J.E. Coupling between arctic feedbacks and changes in poleward energy transport. *Geophys. Res. Lett.* **2011**, *38*, 752. [CrossRef]
11. Overland, J.E.; Wang, M. Large-scale atmospheric circulation changes are associated with the recent loss of arctic sea ice. *Tellus* **2010**, *62*, 1. [CrossRef]
12. Wang, X.; Key, J.R. Arctic surface, cloud, and radiation properties based on the avhrr polar pathfinder dataset. Part ii: Recent trends. *J. Clim.* **2005**, *18*, 2575. [CrossRef]
13. Pithan, F.; Mauritsen, T. Arctic amplification dominated by temperature feedbacks in contemporary climate models. *Nat. Geosci.* **2014**, *7*, 181. [CrossRef]
14. Ding, Q.; Schweiger, A.; L'Heureux, M. Influence of high-latitude atmospheric circulation changes on summertime Arctic sea ice. *Nat. Clim. Chang.* **2017**, *7*. [CrossRef]

15. Duarte, C.M.; Lenton, T.M.; Wadhams, P.; Wassmann, P. Abrupt climate change in the arctic. *Nat. Clim. Chang.* **2012**, *2*, 60. [CrossRef]
16. Schweiger, A.; Lindsay, R.; Zhang, J.; Steele, M.; Stern, H.; Kwok, R. Uncertainty in modeled arctic sea ice volume. *J. Geophys. Res. Oceans* **2011**, *116*, 128. [CrossRef]
17. Overland, J.E.; Wang, M. When will the summer arctic be nearly sea ice free? *Geophys. Res. Lett.* **2013**, *40*, 2097. [CrossRef]
18. Wu, B.; Handorf, D.; Dethloff, K.; Rinke, A.; Hu, A. Winter weather patterns over northern eurasia and arctic sea ice loss. *Mon. Weather Rev.* **2013**, *141*, 3786. [CrossRef]
19. He, J.H.; Wu, F.M.; Qi, L. Decadal/interannual linking between autumn arctic sea ice and following winter eurasian air temperature. *Chin. J. Geophys.* **2015**, *58*, 1089.
20. Mcguire, A.D.; Iii, F.S.C.; Walsh, J.E.; Wirth, C. Integrated regional changes in arctic climate feedbacks: Implications for the global climate system. *Annu. Rev. Environ. Resour.* **2006**, *31*, 61. [CrossRef]
21. Turner, J.; Overland, J.E.; Walsh, J.E. An arctic and antarctic perspective on recent climate change. *Int. J. Climatol.* **2007**, *27*, 277. [CrossRef]
22. Maturilli, M.; Herber, A.; König-Langlo, G. Climatology and time series of surface meteorology in Ny-Ålesund, Svalbard. *Earth. Syst. Sci. Data* **2012**, *5*, 1057. [CrossRef]
23. Nordli, Ø.; Przybylak, R.; Ogilvie, A.E.J.; Isaksen, K. Long-term temperature trends and variability on spitsbergen: The extended svalbard airport temperature series, 1898–2012. *Polar Res.* **2014**, *33*, 91. [CrossRef]
24. Wood, K.R.; Overland, J.E.; Jónsson, T.; Smoliak, B.V. Air temperature variations on the atlantic-arctic boundary since 1802. *Geophys. Res. Lett.* **2010**, *37*, 204. [CrossRef]
25. Rland, E.J.F.; Benestad, R.; Hanssenbauer, I.; Haugen, J.E.; Skaugen, T.E. Temperature and precipitation development at svalbard 1900–2100. *Adv. Meteorol.* **2011**, *2011*, 13.
26. Hanna, E.; Mernild, S.H.; Cappelen, J.; Steffen, K. Recent warming in greenland in a long-term instrumental (1881–2012) climatic context: I. evaluation of surface air temperature records. *Environ. Res. Lett.* **2012**, *7*, 189. [CrossRef]
27. Fyfe, J.C.; Meehl, G.A.; England, M.H.; Mann, M.E.; Santer, B.D.; Flato, G.M. Making sense of the early-2000s warming slowdown. *Nat. Clim. Chang.* **2016**, *6*, 224. [CrossRef]
28. Huang, J.; Zhang, X.; Zhang, Q. Recently amplified arctic warming has contributed to a continual global warming trend. *Nat. Clim. Chang.* **2017**. [CrossRef]
29. Onarheim, I.H.; Smedsrud, L.H.; Ingvaldsen, R.B.; Nilsen, F. Loss of sea ice during winter north of svalbard. *Tellus* **2014**, *66*, 70. [CrossRef]
30. Overland, J.E. Future arctic climate changes: Adaptation and mitigation time scales. *Earths Future* **2014**, *2*, 68. [CrossRef]
31. Zhang, X.; Walsh, J.E.; Zhang, J.; Bhatt, U.S.; Ikeda, M. Climatology and interannual variability of arctic cyclone activity: 1948–2002. *J. Clim.* **2004**, *17*, 2300. [CrossRef]
32. Sepp, M.; Jaagus, J. Changes in the activity and tracks of arctic cyclones. *Clim. Chang.* **2011**, *105*, 577. [CrossRef]
33. Schauer, U.; Fahrbach, E.; Osterhus, S.; Rohardt, G. Arctic warming through the fram strait: Oceanic heat transport from 3 years of measurements. *J. Geophys. Res.* **2004**, *109*, 259. [CrossRef]
34. Polyakov, I.V.; Alekseev, G.V.; Timokhov, L.A.; Bhatt, U.S.; Colony, R.L.; Simmons, H.L. Variability of the intermediate Atlantic water of the arctic ocean over the last 100 years. *J. Clim.* **2010**, *17*, 4485. [CrossRef]
35. Spielhagen, R.F.; Werner, K.; Sørensen, S.A.; Zamelczyk, K.; Kandiano, E.; Budeus, G. Enhanced modern heat transfer to the arctic by warm Atlantic water. *Science* **2011**, *331*, 450. [CrossRef] [PubMed]
36. Sato, K.; Inoue, J.; Watanabe, M. Influence of the Gulf Stream on the Barents Sea ice retreat and Eurasian coldness during early winter. *Environ. Res. Lett.* **2014**, *9*, 084009. [CrossRef]
37. Walczowski, W.; Piechura, J. Influence of the west Spitsbergen current on the local climate. *Int. J. Climatol.* **2011**, *31*, 1088. [CrossRef]
38. Ruggieri, P.; Buizza, R.; Visconti, G. On the link between Barents-Kara sea-ice variability and European blocking. *J. Geophys. Res. Atmos.* **2016**, *121*, 5664. [CrossRef]
39. Francis, J.A.; Hunter, E. Changes in the fabric of the arctic's greenhouse blanket. *Environ. Res. Lett.* **2007**, *2*, 045011. [CrossRef]
40. Rinke, A.; Melsheimer, C.; Dethloff, K.; Heygster, G. Arctic total water vapor: Comparison of regional climate simulations with observations, and simulated decadal trends. *J. Hydrometeorol.* **2009**, *10*, 113. [CrossRef]

41. Serreze, M.C.; Barrett, A.P.; Stroeve, J. Recent changes in tropospheric water vapor over the arctic as assessed from radiosondes and atmospheric reanalyses. *J. Geophys. Res.* **2012**, *117*, 10104. [CrossRef]
42. Maturilli, M.; Kayser, M. Arctic warming, moisture increase and circulation changes observed in the Ny-Ålesund homogenized radiosonde record. *Theor. Appl. Climatol.* **2016**, *130*, 1. [CrossRef]
43. Solomon, S.; Rosenlof, K.H.; Portmann, R.W.; Daniel, J.S.; Davis, S.M.; Sanford, T.J. Contributions of stratospheric water vapor to decadal changes in the rate of global warming. *Science* **2010**, *327*, 1219. [CrossRef] [PubMed]
44. Solomon, S.; Daniel, J.S.; Rd, N.R.; Vernier, J.P.; Dutton, E.G.; Thomason, L.W. The persistently variable "background" stratospheric aerosol layer and global climate change. *Science* **2011**, *333*, 866. [CrossRef] [PubMed]
45. Feng, C.; Wu, B.Y. Enhancement of Winter Arctic Warming by the Siberian High over the Past decade. *Atmos. Ocean. Sci. Lett.* **2015**, *8*, 257.
46. Kosaka, Y.; Xie, S.P. Recent global-warming hiatus tied to equatorial pacific surface cooling. *Nature* **2013**, *501*, 403. [CrossRef] [PubMed]
47. Meehl, G.A.; Arblaster, J.M.; Fasullo, J.T.; Hu, A.; Trenberth, K.E. Model-based evidence of deep-ocean heat uptake during surface-temperature hiatus periods. *Nat. Clim. Chang.* **2011**, *1*, 360. [CrossRef]
48. England, M.H.; Mcgregor, S.; Spence, P.; Meehl, G.A.; Timmermann, A.; Cai, W. Recent intensification of wind-driven circulation in the pacific and the ongoing warming hiatus. *Nat. Clim. Chang.* **2014**, *4*, 222. [CrossRef]
49. Chen, X.; Tung, K.K. Climate varying planetary heat sink led to global-warming slowdown and acceleration. *Science* **2014**, *345*, 897. [CrossRef] [PubMed]
50. Langehaug, H.R.; Matei, D.; Eldevik, T.; Lohmann, K.; Gao, Y. On model differences and skill in predicting sea surface temperature in the Nordic and Barents seas. *Clim. Dyn.* **2017**, *48*, 913. [CrossRef]
51. Årthun, M.; Eldevik, T. On anomalous ocean heat transport toward the arctic and associated climate predictability. *J. Clim.* **2015**, *29*, 151111130840000. [CrossRef]
52. An, W.; Hou, S.; Hu, Y.; Wu, S. Delayed warming hiatus over the Tibetan plateau. *Earth Space Sci.* **2017**, *4*, 128. [CrossRef]
53. Turner, J.; Ju, H.; White, I.; King, J.C.; Phillips, T.; Hosking, S.; Bracegirdle, T.; Marshall, G.J.; Mulvaney, R.; Deb, P. Absence of 21st century warming on Antarctic Peninsula consistent with natural variability. *Nature* **2016**, *535*, 411. [CrossRef] [PubMed]

condensed matter

Article

Synchrotron Radiation Research and Analysis of the Particulate Matter in Deep Ice Cores: An Overview of the Technical Challenges

Giannantonio Cibin [1,*], Augusto Marcelli [2,3], Valter Maggi [4,5], Giovanni Baccolo [4,5], Dariush Hampai [2], Philip E. Robbins [1], Andrea Liedl [2], Claudia Polese [2], Alessandro D'Elia [6], Salvatore Macis [2,7], Antonio Grilli [2] and Agostino Raco [2]

[1] Diamond Light Source, Harwell Science and Innovation Campus, Didcot OX110DE, UK
[2] Istituto Nazionale di Fisica Nucleare, Laboratori Nazionali di Frascati, via Enrico Fermi 40, Frascati, I-00044 Roma, Italy
[3] Rome International Centre for Material Science Superstripes, RICMASS, via dei Sabelli 119A, I-00185 Rome, Italy
[4] Dipartimento di Scienze dell'Ambiente e della Terra, Università degli Studi di Milano Bicocca, Piazza della Scienza, I-20126 Milano, Italy
[5] Istituto Nazionale di Fisica Nucleare, Sezione di Milano-Bicocca, Piazza della Scienza, 2, I-20126 Milano, Italy
[6] Department of Physics, University of Trieste, Via A. Valerio 2, I-34127 Trieste, Italy
[7] Department of Physics, Università Sapienza, Piazzale Aldo Moro 5, I-00185 Rome, Italy
* Correspondence: giannantonio.cibin@diamond.ac.uk; Tel.: +44-1235-778645

Received: 26 April 2019; Accepted: 24 June 2019; Published: 27 June 2019

Abstract: Airborne dust extracted from deep ice core perforations can provide chemical and mineralogical insight into the history of the climate and atmospheric conditions, with unrivalled temporal resolution, time span and richness of information. The availability of material for research and the natural complexity of the particulate, however, pose significant challenges to analytical methods. We present the developments undertaken to optimize the experimental techniques, materials and protocols for synchrotron radiation-based analysis, in particular for the acquisition of combined Synchrotron Radiation X-Ray Fluorescence and X-ray Absorption Spectroscopy data.

Keywords: synchrotron radiation; ice core, atmospheric mineral dust; X-ray absorption spectroscopy

---

1. Introduction

The same phenomena that make deep ice cores precious archives of environmental information are at the source of the technical challenges that studies of chemical, isotopic, mineralogical composition face. This is particularly true where ice cores are used for the acquisition of data aiming at accurate reconstructions of the temporal evolution of atmospheric conditions. Glacier location, atmospheric circulation patterns, ice accumulation rates and the local meteorological conditions determine whether ice preserves, in a time-ordered sequence, samples of the deposited snow. The combined effect of accumulation rate, glacier dynamic, local climatic conditions and geothermal heat flux define the maximum timespan of the preserved ice. The low deposition rates and overall low temperature conditions in Antarctica guarantee, therefore, maximal time span records, but this impacts on the achievable temporal resolution; at the same time, the long distance from sources of airborne particulates and the snow capture mechanisms limit the overall concentration of atmospheric components that precipitate and get stored in ice. Snow deposition rates in Antarctica can be as low as 50 mm equivalent water per year, and this allows deep ice cores to cover past climates back to several hundred ka [1]. Because of the coverage thickness, the accumulation usually leads to unperturbed storage which are less affected by lateral ice movements (when compared e.g., to alpine environments). Also, the provenance of solid material found in ice is atmospheric or meteoric, so Antarctic ice collects non-local information.

Climate and atmospheric information from deep ice cores is first obtained from accurate analysis of ice itself, in particular through the measurement of deuterium and oxygen stable isotope composition, ionic concentrations and gas analysis (e.g., Petit [1]). These parameters provided a direct measurement of temperature across the last few climatic cycles.

Solid (insoluble) particulate represents a significant fraction among the ice core components, as its compositional analysis can reveal temporal variations of the activity of dust sources and of their environmental conditions. Those have consequences on global bio geochemical cycles, and influence mineral alteration under different climatic conditions [2].

Dust content and grain size measurements performed using liquid counter methods [1,3] have indicated that the dust deposition rate also has a close correlation with climate. The rate reflects the evolution of dust sources and of the atmospheric circulation patterns. Comparative analyses with samples from potential source areas, using TIMS (Thermal Ionization Mass Spectrometer) isotopic analysis on the ratios $^{87}Sr/^{86}Sr$ and $^{143}Nd/^{144}Nd$, has indicated a South American provenance of dust deposited in inner East Antarctica [4] in cold periods.

Beyond elemental composition information, obtaining a mineralogical analysis of the particulate would naturally complement these results with the potential of giving much richer insight into both the source and the transport phenomena across the last climatic periods [5–8]

Technical challenges clearly come from the small amounts of solid material available for analysis. The particulate concentration in ice [9,10] in deep ice cores from Antarctic drilling locations, for example, oscillates, depending on the climatic period samples, with dramatic variations from 500 ng/g down to 20 ng/g during glacial and interglacial periods, respectively. These concentrations of particulate in ice, however, are well below the detection limits of modern mineralogical analytical techniques. Extracting significant amounts of the insoluble fraction for chemical and mineralogical analysis therefore requires careful handling and clean protocols; at the same time, it is necessary to minimize the ice core sections used for extraction, considering both the sample rarity (deep ice core drilling present significant challenges involving international collaborations [4]) and the need to preserve the time resolution, directly linked to the ice core section depth profile being sampled. The amounts of particulate extracted will be necessarily minimal. We will introduce here the development in instrumentation and protocols undertaken by our group to optimize the collection of synchrotron radiation X-ray Fluorescence (XRF), X-ray Absorption Spectroscopy (XAS) and X-ray Powder Diffraction (XRD) data on such samples.

2. Methods

2.1. Elemental Composition Analysis

Elemental composition analysis of particulate from deep ice cores has been undertaken before, using Particle-Induced X-ray emission (PIXE) [11,12]. Similar to other X-ray emission detection methods, PIXE analysis provides concentration information for most elements (in practice all elements heavier than Na), has a high sensitivity so is suitable for detecting components in low concentration and is non-destructive for mineral samples. However, PIXE information is limited to elemental analysis, and the exposure to high-energy proton beams has direct effects on the stability of the supports. This possibly precludes re-using the samples in further analyses. As an alternative to proton-induced emission, X-ray based methods can overcome these limitations. Lab-based instrumentation makes it possible to detect trace element concentrations in particular, with the Total Reflection X-Ray Fluorescence configuration, exploiting the advantage that such a configuration minimizes the contribution from sample supports and enhances the signal coming from the materials deposited on flat surfaces. The growing importance of XRF and related techniques is demonstrated by the continuous development and size of its community; advances on materials, technology and procedures are regularly reported in contributions published on *J. Anal. At. Spectrom.* [13]. The use of synchrotron-based instrumentation clearly provides substantial advantages due to the availability of intense, highly collimated and, most importantly, tunable monochromatic beams, at the expense of beam time availability and limited access due to intense competition for access to synchrotron beamlines. Finally, X-ray based analyses have the important characteristics that they don't alter the sample (radiation damage in mineral systems is not as critical as for other sample matrices).Multiple techniques, make it possible to get structural rather than elemental information, and they can be applied on the same sample, provided that sample preparation is compatible with different experimental needs. The increasing importance of synchrotron radiation is evidenced by the increasing list of instruments dedicated to XRF and related methods. A comprehensive list of XRF beamlines is reported in the recent paper from Karydas et al. [14]. It is worth mentioning, however, that recent beamline development efforts are mostly aimed at reducing the beam sizes, due to the scientific potential of microscopic methods e.g., in the environmental and life sciences. Those experiments are particularly demanding in terms of beamtime allocation times, and do not guarantee the efficient use of beam times when average XRF and XAS information is needed, as in our case.

2.2. Sample Preparation and Handling

Dust extraction from ice core samples was undertaken in a clean room environment at the EuroCold Lab facility, University Milano-Bicocca. After the removal with three successive baths in ultra-pure water of the contaminated core surface, ice samples were melted at room temperature in a laminar flow bench and dust was extracted by filtration [5]. All materials used during this procedure are rinsed with ultrapure water and undergo a deep cleaning with hyper-pure nitric acid. The choice of materials and the cleaning procedures have been developed over time, and a detailed analysis of the filter materials using Neutron Activation Analysis has been recently reported in the papers of Baccolo et al. [12]. For our synchrotron experiments dust was deposited on polycarbonate filters with 0.45 μm pore size. Filters were mounted on clean polytetrafluoroethylene holders. The effectiveness of the cleaning protocol was verified on the beamline by comparing XRF measurements on filters as-bought, after the cleaning procedure and filtering the ultrapure water used for container rinsing.

Considering that beamline measurement spaces are not usually prepared to operate in clean-room conditions, we developed a protocol to minimize sample contamination during transport and handling after filtration and during the preparation carried out in the *EuroCold* laboratory [15]. The samples were transported in triple containment, in heat-sealed clean polyethylene bags. Outer and intermediate containments were removed in Diamond clean room and sealed samples transferred to the beamline in clean sealed containers. To allow a clean introduction of samples into the experimental chamber, the latter was extended with an ambient pressure glove box (built in-house), nitrogen-filled, with internal access to the experimental chamber sample load door. The last containment layer was removed just before insertion of the PTFE filter support to the chamber holder with the help of clean tweezers. After measurement, the samples were again sealed in the original envelopes before being returned to the clean chamber for long-term storage. While this handling sequence is probably not critical for experiments programmed in individual batches, it allows repeated measurements to be undertaken in different experimental runs. Considering the intrinsic variability (including seasonal oscillations) of natural dust composition deposited in ice cores, running acquisition campaigns in separate experimental visits has allowed us to refine the sample selection on the basis of previous measurements, and to assess the nature of outliers in the temporal sequence being analyzed. Maintaining clean sample conditions after measurement has therefore allowed cross-checking of the experimental setups to occur, ensuring compatibility of the results across different experiments.

For each experimental run, blank filters were prepared using the same methods as the dust samples, and SR-XRF data were collected to validate the experimental run and ensure that during transport or storage, no significant contamination occurred. Also, a calibration batch of reference standards prepared with comparable amounts of NIST reference soil (NIST 2709a, San Juaquin soil [16]) was regularly measured.

2.3. Experimental

For this work, we had access to two facilities: the Stanford Synchrotron Radiation Lightsource (beamlines 6-2 and 7-2) and Diamond Light Source (beamline B18). In both cases, the beamlines were equipped with double-crystal monochromators, equipped with Si(111) crystals. At SSRL the beam, the liquid-nitrogen cooled monochromators were focused with mirrors in the vertical direction to give a line beam profile of the sample for grazing-incidence data collection, with a beam vertical size of approximately 200 µm. The experimental chamber for total reflection fluorescence and GI-XAS in vertical deflection geometry was developed under the CRYOALP [12] program. Data were acquired using a Silicon Drift detector from KETEK GmbH (Munchen, UK) with Canberra (now Mirion technologies, Meriden, CT, USA) electronics. On B18, the beam was focused horizontally with a FWHM of 200 µm for Total reflection X-Ray Fluorescence and Grazing Incidence X-ray Absorption Spectroscopy acquisitions in the horizontal plane, and defocused to give a 1×1 mm^2 beam footprint on the sample. Membrane samples were aligned for conventional 45° incidence measurements. Data were acquired with a Vortex 4-elements silicon drift detector from SII (Hitachi High Technologies Science America, Nothridge, CA, USA), and Xmap (XIA LLC, Hayward, CA, USA) and XSPRESS3 (Quantum Detectors, Didcot, UK) electronics.

3. Results and Discussion

3.1. X-Ray Fluorescence and X-Ray Absorption Spectroscopy in Grazing Incidence Geometry

When the critical information to be extracted from a sample is coming from a surface, as in the case of particulate deposited on a solid flat substrate, orienting the latter so the incoming beam is in grazing incidence geometry substantially reduces the interaction with the support thanks to X-ray total external reflection. In particular, the substrate does not generate diffuse scattering if its surface roughness is low. The reflection critical angle, which allows the total reflection regime to be established, depends on the substrate composition and X-ray beam energy. This setup takes advantage of a two-fold increase in the amplitude signal obtainable from the creation of a stationary wave on the reflecting surface. Finally, in this geometry, the beam footprint on the support gets significantly increased, allowing for the detection of large surface sections.

We collected our initial set of data on Antarctic particulate samples at SSRL (beamlines 6-2 and 7-2) using a dedicated experimental chamber [17], and at Diamond Light Source (beamline B18). In Figure 1 we compare XRF results from samples deposited on filters and analyzed through normal incidence geometry with the ones deposited on Si wafers and measured in total reflection geometry. The single measurement efficiency and data quality obtainable in total reflection mode are significantly improved, as expected. The total reflection X-ray Fluorescence (TXRF) geometry allows us to probe a large surface area, as the beam footprint on the sample is extended thanks to the $1/\sin(\theta)$ (θ being the grazing incidence angle) geometrical factor; however, several considerations come into play. First, this setup requires dedicated experimental systems, and total reflection setups are not commonly available at XAS beamlines. This means that for long term programs, specialised chambers must be developed and integrated with different synchrotron radiation facilities, and logistics aspects thus have to be considered.

Other limitations of the TXRF method are: the need of focusing the beam down to a few hundred microns; time consumption for the sample alignment, and the importance of keeping the beam position extremely stabile during the measurements, in particular during XAS scans. In addition, the grazing incidence angle must be adjusted with high accuracy, as the fluorescence line intensity of elements present in the support is critically dependent on the angle. This is clearly essential in our case, as the XRF analysis must include an accurate quantification of Silicon, which is present in the common clean materials used as supports for TXRF analysis (commonly used are quartz and Si wafers). Alternative materials to avoid interference with Si and most of the elements of interest for natural particulate analysis are in practice not available; in particular, we found that plastic-based supports with had contaminations, inferior surface quality and the total reflection angles were significantly lower than Si. Significantly heavier elements present lower energy absorption edges (L and M) that always fall into the region of interest for major element analysis in natural systems. The deposition of particulate on substrates requires additional manipulation steps, such as removal from filters with ultrapure water and re-deposition on substrates, with additional risks of contamination. This deposition method causes a visible "coffee stain" pattern on the substrate, which could lead to a fractionation of different components leading to different results, depending on the part of the sample illuminated by the beam [17].

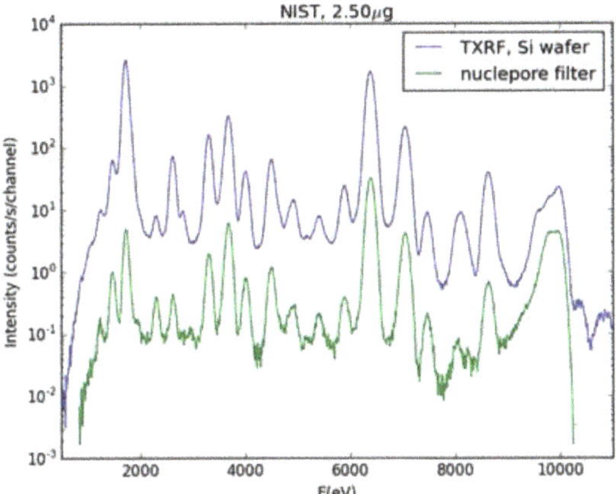

Figure 1. Comparison between TXRF and normal incidence XRF acquired on reference NIST soil samples, with same amounts deposited. Apart from the overall higher intensity, the TXRF measurement displays a significantly different Si signal intensity due to the substrate contribution, and at the same time a lower elastic scattering contribution to the spectrum. The remaining signal from other elements is consistent across the two measurements, indicating the background contribution from filter contamination is negligible.

3.2. Normal Incidence Geometry

On the opposite side, the use of polymer membrane filters presents issues coming from the nature of the material and the experimental arrangements. The incoming beam probes the whole support thickness, so composition contamination of the filters will be more important than in total reflection geometry (Figure 2 reports a comparison between XRF acquired on two blank filters—PMMA and metacrylate—and a standard soil sample prepared depositing the smallest amount used in our research). Minimizing the filter thickness is important as well. This does not only help with the direct contamination signal, but helps reducing the overall diffuse beam scattering (a limiting factor for fluorescence detectors for dilute components analysis). It is expected that most of the fluorescence signal also from the low Z components should come from the front filter surface, so corrections for the low energy efficiency are considered minor.

Regarding the sample preparation, filtering does not pose the same problems regarding potential separation between components due to coffee stain. However, if the sample is deposited approximately on the same surface (as with the use of common lab filtering systems, with a deposition diameter of 13 mm), the sample density "seen" by the beam will be significantly lower, as discussed.

At the same time, filtration on samples characterized by small particle numbers does not guarantee a homogeneous deposition. In Figure 3, we present the composition profile acquired on a filter prepared with this method. It is clear that significant variations across the sample diameter are present, i.e., up to 50% for the Ca/Fe ratio.

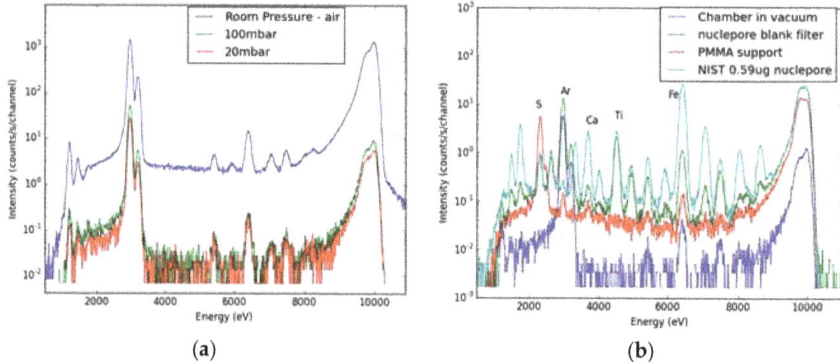

Figure 2. (**a**) Variation of background signal in vacuum and atmospheric pressure conditions from the experimental chamber. The atmospheric pressure measurement shows, beyond the strong contribution of Ar and direct scattering, a strong enhancement of signals coming from fluorescence from steel components present in the experimental chamber materials, excited by the diffuse scattering. Visible lines are Ar Kα and Kβ emission lines (main and escape peaks at low energy), Cr, Fe and Ni. (**b**) Comparison between the experimental chamber background, spectra from one of many the polymer supports considered (PMMA, polymethyl metacrilate) showing traces of sulfur contamination, a *Nuclepore* polycarbonate blank filter before acid cleaning and a dilute NIST 2709a sample, consisting in 0.59 µg mineral particles deposited on *Nuclepore* filter.

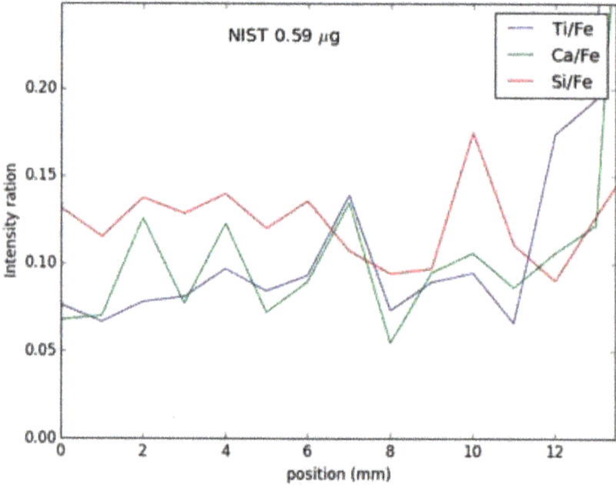

Figure 3. Intensity ratios between different elemental components extracted from XRF spectra taken at different vertical positions on a low concentration (0.59 µg/cm^2) sample, deposited on nuclepore filtersover a 13 mm diameter area. The total beam footprint on the sample was adjusted to match the experimental step (1 × 1 mm^2). The variability in the concentration ratios highlights that potential composition variation artifacts could be introduced if the sample deposition happens on large surfaces.

This can be moderated with a reduction of the deposition area, and a change of the deposition method to concentrate the sample in the smallest possible area of the filter surface. An additional step in this direction has been undertaken by considering automated deposition systems [18], which proved to be efficient in minimizing the amount of ice necessary to reach a desired sample concentration on small areas (few mm^2) that are directly comparable with the beam sizes attainable on modern focused beamlines. This deposition could dramatically improve the homogeneity of the deposited material and, at the same time, reduce both the deposition time and sample consumption, and increase the local concentration of the sample on the filter. For XRF analysis, a limiting factor for reducing the detection limit is the presence of an experimental background signal. Several aspects affecting the signal quality were taken into consideration. The general source for the background was found to be diffuse scattering, causing secondary emission from the experimental chamber wall materials, while contamination coming from the filters' support was dealt with by using filter supports and sample holders all manufactured in PTFE. Reduction of the scattering intensity and minimization of the background signal has been obtained by:

(a)   Reducing the low level, diffuse, out-of-focus contribution from the beamline optics (caused by mirror coatings roughness and vacuum windows) using beam stripping slits placed just before the ionization chamber monitoring the incoming beam intensity;

(b)   Reducing further contributions from small angle scattering by the ionization chamber windows (*Kapton*, thickness: 50 µm), by modifying the position of the ion chamber, increasing its distance to the entrance of the experimental chamber and adding a plastic collimator at the entrance of the chamber. The collimator, while still allowing for scanning the experimental chamber to map the sample position, ensured any diffuse beam to propagate through the sample position to the end aperture of the vacuum chamber (Figure 4);

(c)   Reducing diffuse scattering from the residual gas in the vacuum chamber, improving the pumping system (experiments were always run under turbo-molecular pumping, ensuring a measurement pressure lower than 1 mbar);

(d)   Filtering the fluorescence from the chamber walls by covering the internal walls with a 2 mm-thick polycarbonate lining and ensuring the beam exit path is free from metal sections close to the diffuse scattering from the sample support;

(e)   Finally, minimizing the collection field of view by adding a clean PTFE collimator placed in front of the 4-elements SDD detector.

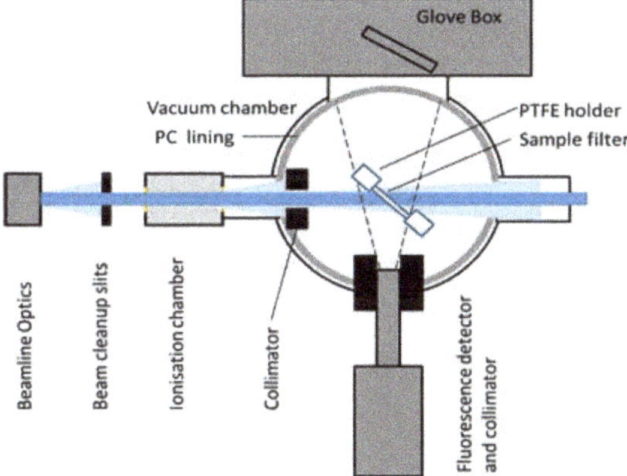

Figure 4. Layout of the experimental setup. In black are indicated the beam collimation components (cleanup slits, experimental chamber and fluorescence detector collimator), reducing the spatial distribution of beam diffused by optical elements and windows (in light blue). In light grey, the vacuum chamber lining in polycarbonate used to minimize contributions of X-ray fluorescence from the experimental chamber, excited from beam scattered from the filter and sample support.

3.3. X-Ray Absorption Spectroscopy

The optimization of the experimental setup for XRF measurements allowed us to collect, in an efficient way, a full set of X-ray Absorption Spectroscopy data at the edge of elements of interest to obtain the information on oxidation states and element-specific coordination that XAS provides to support, in particular, through direct comparison with relevant standards, the identification of the mineral composition of our samples [11]. In particular, we were able to collect data at the Fe, Ti and Ca K edges on a significant set of samples, across the last climatic transition, during a long-term proposal on B18 (Figure 5). These elements are important in the case of ice core particulate in Antarctica, as information on Fe speciation can lead to an identification of the dust sources. The relationship between Ti and Fe chemistry brings information on local vs. long range transport in Antarctica, and Ca can be of continental or marine origin. The samples' concentration was optimized to give a significant total photon count rate on the fluorescence detector, which was limited to approximately 200 kcps/element over all the project run to ensure all samples data were acquired in the same experimental conditions (in particular, the same maximum detector dead time fraction which is dependent on the total count rate). The total signal from polycarbonate filters at the energies corresponding to the Fe K edge absorption edge measurements (up to 7.8 keV) was found to always be in the ideal detector range, with samples containing total amounts of particulate in the range of ~3–17 µg, distributed over an approximate surface of 0.25 cm². It is worth noting that this indicates (at least for the Fe K edge) that no significant advantages would be attainable from the use of more intense sources, such as wiggler/undulator based beamlines, as the experiment duration would be detector-limited. An optimal acquisition time efficiency was obtained using a continuous energy scanning mechanism—already developed on B18—so minimal dead times during acquisitions were present. For the Fe K edge, a total acquisition time of up to ~15 min was sufficient for extended XANES spectra (data were collected usually up to 7500 eV to ensure accurate normalization, and up to 7800 eV for a reduced number of concentrated samples to allow for EXAFS evaluation).

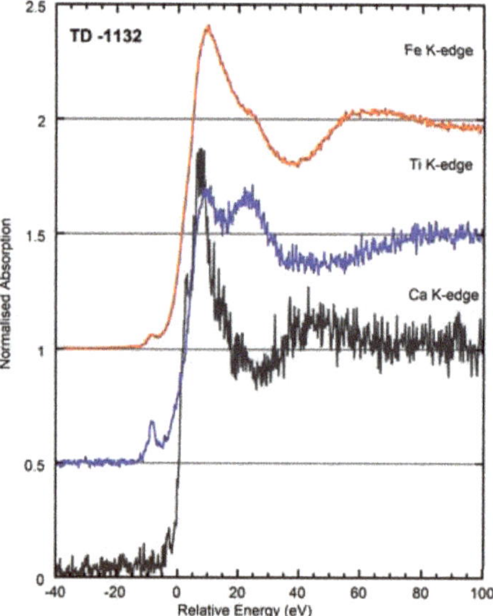

Figure 5. X-ray Absorption spectra collected at the Fe, Ti and Ca K-edge on a representative sample (1.2 µg/cm² of Antarctic dust on a polycarbonate membrane) from Talos Dome. The energy axis is relative to respective absorption edges. The total acquisition time for Fe K-edge was 12 min, while for Ti and Ca we collected data for 30 and 45 min, respectively.

Using continuous scans [19], we could collect several measurement repetitions over short time periods (approximately 3 min/scan), which allowed us to monitor potential radiation-induced effects such as photoreduction/oxidation, and to adjust the total acquisition time per sample to obtain a homogeneous data quality across the whole sample set. With the SDD detector placed at ~50 mm from the sample surface, it was possible during the same experimental session to acquire XANES data from lower energy edges (Ca and Ti), with sufficient quality to allow for a good recognition of the coordination environment of those two elements. It is important to note that the signal drops rapidly at low energies, because of the lower fluorescence yield, so acquisition times were proportionally longer (we limited our collection time to 1 h/sample).

This setup allowed us to collect a full set of XRF and XAS data on the whole timespan covered by the Talos Dome perforation, located on the East Antarctic Ice Sheet [20]. This is a set of over 200 XRF and XAS individual acquisitions, which includes a significant fraction of data taken on background, blanks, and repetition measurements made necessary by the need to ensure that reproducible experimental conditions are met for all the experiments, split across several visits over a long-term project at Diamond Light Source. The first experiments were dedicated exclusively to the development and optimization of the acquisition and sample preparation, and determined, for example, the strategy of acquiring only data from filter samples, excluding the route of grazing-incidence measurements given the experimental uncertainties evidenced in the first runs.

4. Conclusions

The analysis of the insoluble fraction deposited in deep ice cores requires the development of techniques and protocols which are specific to this field. The small amounts of material available for analysis represent only one of the aspects to be taken into consideration when planning experimental campaigns, in particular if they rely on instrumentation available at large scale facilities. Instrumentation development in close collaboration with the facility staff substantially reduces the detection limits of the techniques, but careful sample preparation must be taken into consideration from the beginning, and must be an integral part of the experimental plans. The facts that access to beamtime is restricted and experimental campaigns could be divided in successive experimental visits, which are necessary to collect enough experimental points in the timescale covered by deep ice core to give statistically significant results, challenge the reproducibility of the instrumentation and of the preparation procedure, and require careful cross-checking and validation. We achieved a robust analysis set of consistent data on deep ice cores by incrementally developing these methods, thanks to a long term and close collaboration of all the parties involved, both from the scientific and technical sides, and by dedicating significant part of the beamtime to the analysis of instrument performance via long-term experimental campaigns.

Author Contributions: conceptualization, A.M., G.C., V.M.; methodology, A.M., D.H., G.C., G.B., P.E.R., A.R., A.G., S.M.; investigation, A.M., D.H., G.C., G.B., A.L., C.P., S.M.; software, G.C., D.H., G.B.; resources, A.M., V.M., G.B., G.C.; formal analysis, G.C., G.B., D.H., C.P., A.L.; data curation, G.C., D.H., C.P., G.B., A.L.; writing-original draft preparation, G.C.; writing-review and editing, all authors; supervision, project administration, funding acquisition, A.M., V.M.

Funding: Part of the results presented were acquired at the Stanford Synchrotron Radiation Lightsource, SLAC National Accelerator Laboratory, supported by the U.S. Department of Energy, Office of Science, Office of Basic Energy Sciences under Contract No. DE-AC02-76SF00515. Part of the work is done using specific funding by *Dipartimento Affari Regionali e Autonomie* (DARA) of the *Italian Presidenza del Consiglio dei Ministri*.

Acknowledgments: We are grateful for the support of SSRL staff, in particular Piero Pianetta and Matthew Latimer for helping with the beamline setup at 6-2 and 7-2. The authors wish to acknowledge Diamond Light Source for provision of beamtime within proposals sp7314, sp8372 and sp9050. We sincerely acknowledge Annibale Mottana for having triggered these researches and for his continuous support.

Conflicts of Interest: The authors declare no conflict of interest.

References

1. Petit, J.R.; Jouzel, J.; Raynaud, D.; Barkov, N.I.; Barnola, J.M.; Basile, I.; Bender, M.; Chappellaz, J.; Davis, M.; Delaygue, G. Climate and Atmospheric History of the Past 420,000 Years from the Vostok Ice Core, Antarctica. *Nature* **1999**, *399*, 429–436. [CrossRef]
2. Maher, B.; Prospero, J.M.; Mackie, D.; Gaiero, D.; Hesse, P.P.; Balkanski, Y. Global connections between aeolian dust, climate and ocean biogeochemistry at the present day and at the last glacial maximum. *Earth Sci. Rev.* **2010**, *99*, 61–97. [CrossRef]
3. Delmonte, B.; Andersson, P.S.; Hansson, M.; Schoeberg, H.; Petit, J.; Basile-Doelsch, I.; Maggi, V. Aeolian dust in East Antarctica (EPICA-Dome C and Vostok): Provenance during glacial ages over the last 800 kyr. *Geophys. Res. Lett.* **2008**, *35*, L07703. [CrossRef]
4. EPICA community members. Eight glacial cycles from an Antarctic ice core. *Nature* **2004**, *429*, 623–628. [CrossRef] [PubMed]
5. Sala, M.; Sala, M.; Delmonte, B.; Frezzotti, M.; Proposito, M.; Scarchilli, C.; Maggi, V.; Artioli, G.; Daoiaggi, M.; Marino, F.; et al. Evidence of calcium carbonates in coastal (Talos Dome and Ross Sea area) East Antarctica snow and firn: Environmental and climatic implications. *Earth Planet. Sci. Lett.* **2008**, *271*, 43–52. [CrossRef]
6. Briat, M.; Royer, A.; Petit, J.R.; Lorius, C. Late glacial input of eolian continental dust in the Dome C ice core: Additional evidence from individual microparticle analysis. *Ann. Glaciol.* **1982**, *3*, 27–30. [CrossRef]
7. Gaudichet, A.; Angelis, M.D.; Lefevre, R.; Petit, J.R.; Korotkevitch, Y.S.; Petrov, V.N. Mineralogy of insoluble particles in the Vostok Antarctic ice core over the last climatic cycle (150 kyr). *Geophys. Res. Lett.* **1988**, *15*, 1471–1474. [CrossRef]
8. Dapiaggi, M.; Sala, M.; Artioli, G.; Fransen, M.J. Evaluation of the phase detection limit on filter-deposited dust particles from Antarctic ice cores. *Zeit. Kristallog. Kristallogr. Suppl.* **2007**, *26*, 73–78. [CrossRef]
9. Lambert, F.; Delmonte, B.; Petit, J.R.; Bigler, M.; Kaufmann, P.R.; Hutterli, M.A.; Stocker, T.F.; Ruth, U.; Steffensen, J.P.; Maggi, V. Dust-climate couplings over the past 800,000 years from the EPICA Dome C ice core. *Nature* **2008**, *452*, 616–619. [CrossRef] [PubMed]
10. Wegner, A.; Fischer, H.; Delmonte, B.; Petit, J.R.; Erhardt, T.; Ruth, U.; Svensson, A.; Vinther, B.; Miller, H. The role of seasonality of mineral dust concentration and size on glacial/interglacial dust changes in the EPICA Dronning Maud Land ice core. *J. Geophys. Res. Atmos.* **2015**, *120*, 9916–9931. [CrossRef]
11. Marino, F.; Maggi, V.; Delmonte, B.; Ghermandi, G.; Petit, J. Elemental composition (Si, Fe, Ti) of atmospheric dust over the last 220 kyr from the EPICA ice core (Dome C, Antarctica). *Ann. Glaciol.* **2004**, *39*, 110–118. [CrossRef]
12. Baccolo, G.; Maffezzoli, N.; Clemenza, M.; Delmonte, B.; Prata, M.; Salvini, A.; Maggi, V.; Previtali, E. Low background neutron activation analysis: A powerful tool for atmopsheric mineral dust analyisis in ice cores. *J. Radioanal. Nucl. Chem.* **2015**, *306*, 589–597. [CrossRef]
13. Vanhoof, C.; Bacon, J.R.; Ellis, A.T.; Vincze, L.; Wobrauschek, P. 2018 atomic spectrometry update—A review of advances in X-ray fluorescence spectrometry and its special applications. *J. Anal. At. Spectrom.* **2018**, *33*, 1413–1431. [CrossRef]
14. Karydas, A.G.; Czyzycki, M.; Leani, J.J.; Migliori, A.; Osan, J.; Bogovac, M.; Wrobel, P.; Vakula, N.; Padilla-Alvarez, R.; Menk, R.H.; et al. An IAEA multi-technique X-ray spectrometry endstation at Elettra Sincrotrone Trieste: benchmarking results and interdisciplinary applications. *J. Synchrotron Radiat.* **2018**, *25*, 189–203. [CrossRef] [PubMed]
15. EuroCold Lab. Available online: http://www.eurocold.unimib.it/ (accessed on 1 June 2019).
16. Mackey, E.A.; Christopher, S.J.; Lindstrom, R.M.; Long, S.E.; Marlow, A.F.; Murphy, K.E.; Paul, R.L.; Popelka-Filcoff, R.S.; Rabb, S.A.; Sieber, J.R.; et al. *Certification of Three NIST Renewal Soil Standard Reference Materials for Element Content: SRM 2709a San Joaquin Soil, SRM 2710a Montana Soil I, and SRM 2711a Montana Soil II*; NIST Special Publication 260-172; National Institute of Standards and Technology: Gaithersburg, MD, USA, 2010; p. 39. [CrossRef]
17. Cibin, G.; Marcelli, A.; Maggi, V.; Sala, M.; Marino, F.; Delmonte, B.; Albani, S.; Pignotti, S. First combined total reflection X-ray fluorescence and grazing incidence X-ray absorption spectroscopy characterization of aeolian dust archived in Antarctica and Alpine deep ice cores. *Spectrochim. Acta Part B At. Spectrosc.* **2008**, *63*, 1503–1510. [CrossRef]

18. Macis, S.; Cibin, G.; Maggi, V.; Baccolo, G.; Hampai, D.; Delmonte, B.; D'Elia, A.; Marcelli, A. Microdrop Deposition Technique: Preparation and Characterization of Diluted Suspended Particulate Samples. *Condens. Matter* **2018**, *3*, 21. [CrossRef]
19. Dent, A.J.; Cibin, G.; Ramos, S.; Parry, S.A.; Gianolio, D.; Smith, A.D.; Scott, S.M.; Varandas, L.; Patel, S.; Pearson, M.R.; et al. Performance of B18, the Core EXAFS Bending Magnet beamline at Diamond. *J. Phys. Conf. Ser.* **2013**, *430*, 012023. [CrossRef]
20. Maggi, V.; Baccolo, G.; Cibin, G.; Delmonte, B.; Hampai, D.; Marcelli, A. XANES Iron Geochemistry in the Mineral Dust of the Talos Dome Ice Core (Antarctica) and the Southern Hemisphere Potential Source Areas. *Condens. Matter* **2018**, *3*, 45. [CrossRef]

Review

Perspectives of XRF and XANES Applications in Cryospheric Sciences Using Chinese SR Facilities

Wei Xu [1,*], **Zhiheng Du [2,*]**, **Shiwei Liu [2]**, **Yingcai Zhu [1]**, **Cunde Xiao [3]** and **Augusto Marcelli [4,5]**

[1] Beijing Synchrotron Radiation Facility, Institute of High Energy Physics, Chinese Academy of Sciences, Beijing 100049, China; yingcaizhu@ihep.ac.cn

[2] State Key Laboratory of Cryospheric Science, Northwest Institute of Eco-Environment and Resources, Chinese Academy of Sciences, Lanzhou 730000, China; liushiwei1990@lzb.ac.cn

[3] State Key Laboratory of Land Surface Processes and Resource Ecology, Beijing Normal University, 19 Xinjiekouwai Street, Beijing 100875, China; cdxiao@bnu.edu.cn

[4] INFN-Laboratori Nazionali di Frascati, Via E. Fermi 40, 00044 Frascati (RM), Italy; marcelli@lnf.infn.it

[5] RICMASS, Rome International Center for Materials Science Superstripes, Via dei Sabelli 119A, 00185 Rome, Italy

* Correspondence: xuw@mail.ihep.ac.cn (W.X.); duzhiheng10@163.com (Z.D.); Tel.: +86-010-8823-5156 (W.X.)

Received: 6 September 2018; Accepted: 29 September 2018; Published: 8 October 2018

Abstract: As an important part of the climate system, the cryosphere, can be studied with a variety of techniques based on laboratory-based or field-portable equipment in order to accumulate data for a better understanding of this portion of the Earth's surface. The advent of synchrotron radiation (SR) facilities as large scientific interdisciplinary infrastructures has reshaped the scenario of these investigations and, in particular, of condensed matters researches. Many spectroscopic methods allow for characterizing the structure or electronic structure of samples, while the scattering/diffraction methods enable the determination of crystalline structures of either organic or inorganic systems. Moreover, imaging methods offer an unprecedented spatial resolution of samples, revealing their inner structure and morphology. In this contribution, we briefly introduce the SR facilities now available in mainland China, and the perspectives of SR-based methods suitable to investigate ice, snow, aerosols, dust, and other samples of cryospheric origin from deep ice cores, permafrost, filters, etc. The goal is to deepen the understanding in cryospheric sciences through an increased collaboration between the synchrotron radiation community and the scientists working in polar areas or involved in correlated environmental problems.

Keywords: synchrotron radiation; X-ray fluorescence spectroscopy; X-ray absorption fine structure spectroscopy; trace elements; cryospheric sciences; snow; ice; dust

---

1. Introduction

The cryosphere was proposed as the fifth earth sphere alongside the atmosphere, hydrosphere, lithosphere, and biosphere in the ecosystem. Moreover, changes in the cryosphere may reflect global climate changes [1]. For instance, environmental water resources [2] or the pollutant mercury [3] have recently been extensively investigated in China in order to understand global climatic changes and other environmental issues.

Large synchrotron radiation facilities based on dedicated accelerators have been available to users since the early 1990s, and gained an increased interest in the new millennium. After almost six decades of operation, the scenario of synchrotron radiation is going to evolve into low-emittance and quasi-diffraction limit storage rings, either by upgrading the existing facilities (e.g., APS, ESRF, SPring-8, Diamond, Elettra, ALS, etc.) or by building new dedicated storage rings (such as HEPS in

China, SIRIUS in Brazil, MAX IV in Sweden, etc.) [4] At the same time, fourth generation sources, X-ray free electron lasers (FELs), have also been built in Europe, Japan, South Korea, the United States, and in China. These radiation sources offer novel capabilities to investigate the different states of matter from the atomic to nano- and micro-scopic scale.

The continuous increase of synchrotron radiation applications both at storage rings and FELs, in all scientific areas, such as biology, chemistry, materials science, physics, geological and environmental sciences, paleontology, cultural heritage, as well as industrial applications, were made possible by many new instrumental capabilities. In the past, user communities in each scientific area have been established by starting just one or few synchrotron radiation experimental methods. However, many opportunities still remain unexplored in different fields at SR or FEL facilities. Meanwhile, unlike bench top experimental devices, beamlines at large synchrotron radiation facilities are characterized by an Open Access policy, based on the free competition among proposals/experiments that are ranked only by looking at their scientific content at the international level. Actually, the high demand of these facilities restricts their access only to a limited number of researchers, typically running one experiment per year. Only highly competitive proposals are approved and get beamtime. Hence, it is important for each community to know the characteristics of the instruments available at the synchrotron radiation facilities in order to be competitive and successful to solve their open problems.

Cryospheric science is one of the research areas that requires these powerful sources, but at present, only a limited number of scientific researchers have had opportunities to use these infrastructures [5–9]. In this work, after a brief introduction of the various SR techniques that might be of interest for cryospheric researchers, we will address the main techniques that can be performed at Chinese synchrotron radiation facilities in this area, which is of great interest to the Chinese research communities involved in geological, environmental, climatic, and atmospheric research.

2. Synchrotron Radiation-Based Techniques

2.1. X-ray Fluorescence (XRF)-Trace Element Identification

2.1.1. Basics

X-ray fluorescence spectroscopy is based on the emission of secondary photons upon the excitation of core electrons followed by electronic transitions from higher levels to the vacancy state. In atomic physics, K, L, M, and N stand for the main shells numbered as 1, 2, 3, and 4, respectively, while the s, p, d, and f orbitals correspond to different subshells. In X-ray physics, the electron orbitals surrounding the atomic core are indexed as K, L, M, and N, while the K stands for 1s (1/2) shell, L_1 for 2s (1/2), L_2 for 2p (1/2), and L_3 for 2p (3/2). As shown in Figure 1, the X-ray fluorescence line emitted with the transition from the L to K shell is called K_α, while that for the M to K shell is called K_β. For the subshell transitions, the emission lines are named $K_{\alpha 1}$, $K_{\alpha 2}$, $K_{\beta 1}$, $K_{\beta 3}$, and so on. The M to L shell transitions are named as L_α, L_β, L_γ, and so on.

The X-ray fluorescence (XRF) lines of each element are defined by the transition energy among the different orbitals. Each line occurs at a fixed energy and the fluorescence lines are clearly separated for the different elements. The energy value of each emission line allows for identifying the element in the investigated sample. For instance, the K_α and K_β lines of iron can be detected at 6.4 keV and 7.06 keV, respectively. Meanwhile, the L_α, L_β, and L_γ lines of lead occur at 10.6 keV, 12.6 keV, and 14.8 keV, respectively. Usually, one should rely on at least two lines to verify the presence of the corresponding element, because the poor energy resolution XRF lines of some elements may overlap and cannot be distinguished using a conventional detection scheme. For instance, the K_α line of arsenic is quite close to the L_α line of lead. With an energy dispersive or a wavelength dispersive detection scheme, it is possible to distinguish the XRF lines with a high energy resolution.

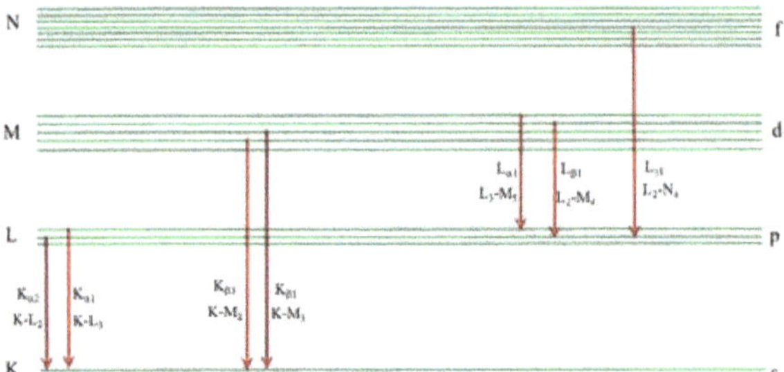

Figure 1. The nomenclature of the X-ray fluorescence lines based on the electron orbital diagram.

2.1.2. Synchrotron Radiation X-ray Fluorescence

A typical synchrotron radiation XRF experimental setup consists of an intense radiation source (bending magnets or an insertion device like a wiggler or an undulator), a transport line that deflects and focuses the X-ray beams, and the end-station where experiments can be performed. Usually, SR-XRF provides a tiny focused beam with a spot size ranging from tens of nanometers up to tens of micrometers, depending on the optical configurations. Some beamlines may offer a spot of variable sizes to fulfill the different experimental demands [10]. In the experimental station, the focused beam is delivered onto the sample and excited fluorescence photons are emitted. To eliminate the strong Compton scattering contribution, the detection geometry has the detector set at 90°, while the sample is set at 45° with respect to the incident beam.

For X-ray fluorescence measurements, one should be concerned about photons illuminating the samples, which will affect the detection limit and the measuring time. There are two classes of experiments that require intense and brilliant synchrotron radiation sources. The first looks at the reconstruction of the spatial heterogeneity in the sample under investigation, and in particular, the mapping of the spatial distribution of the chemical elements. For this type of measurement, we need to carefully evaluate the time necessary to scan the sample area according to spatial resolution and signal to noise (S/N) ratio. For instance, a map 1×1 mm^2 can take several hours using a 5 μm spot with 1 s of acquisition at each point. Another class of SR-XRF experiments looks at elements in trace, that is, at a low and ultralow detection limit, which could be as low as the μg/g level. It should be noted that the absolute detection limit in the mass can reach *fg* and possibly *ag*. These detection values are now possible only at synchrotron radiation beamlines, and the advantage of measuring trace elements with this technique is that it is both highly sensitive and non-destructive. It allows for further analytical techniques to be performed on the same sample. Meanwhile, it should be noted that the total reflection XRF (TXRF) technique that works in the grazing incidence geometry is very well suited for trace-element studies in aqueous solutions, and was successfully applied to investigate the aeolian dust in Antarctica and Alpine deep ice cores [7].

Presently, we have two powerful beamlines at Chinese synchrotron facilities where XRF experiments can be performed, namely: the 4W1B beamline at the Beijing Synchrotron Radiation Facility (BSRF) and the BL15U at the Shanghai Synchrotron Radiation Facility (SSRF) [11]. The first works in the wide bandwidth mode using a double monolayer monochromator, while the second works with the narrow bandwidth mode using a double crystal monochromator. The main working energy at 4W1B is 15 keV, while the X-ray beam is focused down to 20–50 μm in diameter by using a polycapillary lens. However, the energy can be tuned in the range of 5–20 keV and the X-ray beam can be focused down to 2×2 μm^2 by using a couple of Kirkpatrick-Baez mirrors. Users can select the corresponding beamline according to their specific purpose and requirements. The XRF spectra can be fitted to a obtain quantitative evaluation of the elemental concentration by means of calibration curves, using certified standards. The software package PyMCA [12], developed by the European Synchrotron Radiation Facility, is available and is commonly used for data reduction and analysis.

2.2. X-ray Absorption Spectroscopy (XAS)-Atomic Structural Information

In X-ray spectroscopy (XAS), photons excite bound electrons in the atomic core levels (core electrons) to the unoccupied states in the continuum above the Fermi level. The atomic absorption cross-section shows typical step-like jumps at the energy of the core electrons' binding energy. The first interpretation of the modulation of the atomic cross-section over about the 1000 eV photon energy range beyond the absorption edge, was proposed by R.L. Kronig [13], and it was confirmed by using the synchrotron radiation [14], which provides a unique continuum spectrum from the soft to the hard X-ray range. These spectral features are called the "Kronig structure" [13], and since 1970, have used the acronym EXAFS (extended X-ray absorption fine structure) [14]. In this regime, the oscillation of the absorption cross-section is due to the scattering of the fast photoelectron by neighbor atoms confined within the single scattering regime, because of the weak backscattering probability for a fast, high energy photoelectron. The strong sharp absorption peaks appearing in the first hundred eV energy range beyond the absorption edge have been the object of long standing discussions and controversy, generating significant confusion in the literature, as in this energy range the photoelectron energy is small, and therefore the excited photoelectron has a very large scattering cross-section interacting with the neighbor atoms, and its wave-length is much larger than the interatomic distances, thus the single scattering EXAFS approximation is not valid. Moreover, the lifetime of the excited photoelectron is short and its mean free path is also limited, therefore the final short living states are confined to the nanoscale. In 1974, it was proposed that the strong absorption peaks near the absorption edge are due to quasi stationary states of the excited photoelectron degenerate with the continuum localized by strong multiple scattering [15], and it was confirmed experimentally first in disordered aluminum oxide [16], nitrogen gas [17], biological matter [18], surface oxides [19], and complex transition metal oxides, where the acronym XANES (X-ray absorption near edge structure) was proposed [20], in order to indicate the final states localized on the nanoscale by multiple scattering by the atoms surrounding the photo-absorber.

XANES [14] is a widely used probe of both the local structure at an atomic scale, and the electronic structure. X-rays excite core-level electrons and the photoelectrons are scattered by the atoms surrounding the photo-absorber. The scattering of the excited electrons allows for local structural information, such as electronic and geometric structures, to be extracted. XRF and XAS use the same nomenclature, for example, the K-shell electron is excited (K absorption edge appears in XAS spectrum), then K_{α} or K_{β} are the characteristic emission lines in the XRF spectrum. It is worth mentioning that the XAS spectra can be measured by the transmission or fluorescence mode. For the fluorescence detection mode, one should measure the emitted fluorescence lines by scanning the incident energy across the absorption edge. To interpret XAS data, one has to separate the entire spectrum into two regions, namely: the first from 50 eV before up to 60–80 eV (depending on the system and also coordination and bond distances) above the absorption edge, which is called XANES (X-ray absorption near-edge spectroscopy); the second after the XANES region up to 1000 eV or more (depending on the S/N ratio), which is called the EXAFS (extended X-ray absorption fine structure spectroscopy) region [21]. Although one can get both the XANES and EXAFS spectra in one measurement, the interpretation of the XANES and EXAFS spectra are different, and they contain different information. For the XANES spectra, one can obtain the electronic structure and the local geometrical structure (i.e., coordination and bond angles) surrounding the selected absorber atom. By contrast, EXAFS is extremely useful for obtaining the coordination number, the bond distance, and the mean-square-relative displacement of the bonded atoms, through model-based fits. Although both XANES and EXAFS can be interpreted using the multiple scattering theory [10,11], the first is based only on the multiple scattering (MS) analysis in the real space, while the second mainly probes single scattering (SS) events. The EXAFS interpretation requires structural fitting, while the XANES spectra, in addition to simple fingerprinting, can be successfully used to obtain quantitative structural information based on the fitting [22]. Several software packages for the data analysis and interpretations are now available, namely:

(1)   IFEFFIT and Demeter package for data reduction and EXAFS fittings [23];
(2)   Full multiple scattering calculations implemented in FEFF [24] and FDMNES [25] are commonly employed for simulating XANES spectra; and
(3)   XANES fitting as implemented in MXAN [22,26], is based on the comparison between the experimental spectrum and several theoretical calculations generated by changing the relevant geometrical parameters of the site by calculating each configuration with full multiple scattering theory.

2.3. Other Synchrotron Radiation Techniques and Beyond

In addition to XRF and XAFS, it is important to mention that other techniques available at synchrotron radiation facilities may offer important opportunities for cryospheric and environmental researches. Nowadays, several high-energy resolution fluorescence detection mode XANES facilities are available and allow one to measure the near-edge region with more detail [27]. Also, the X-ray diffraction method, a well-established technique that reveals the crystalline structure of organic and inorganic samples with a long-range structural order is available at SR facilities. SR-XRD offers a much faster and more accurate structural refinement than any conventional source on samples of extremely small dimensions (down to micrometer size), so even the analysis of single grains or particulate matter (PM) is possible [28]. Finally, the X-ray tomography can be also used to collect unique 3D images of the internal structure of a sample with an extremely high spatial resolution and a high absorption or phase contrast [29,30]. In the next section, we will briefly illustrate some applications of the different techniques used in environmental sciences, which are of great interest to cryospheric researches.

3. Experimental Applications

3.1. Spectroscopic Methods

Mercury is a global pollutant whose environmental importance steams from its extreme mobility and toxicity. In order to understand its biogeochemical cycle, also in relation to human disturbances, it is important to assess the emitting sources; the dynamics, which govern its transport through the atmosphere; and the deposition on continents and oceans. Primary anthropogenic Hg emissions greatly exceed natural geogenic and biogenic sources, resulting in a massive disturbance of its cycle on a global scale [31]. Moreover, the toxicity is one of the main concerns in biology, and the chemical speciation is extremely useful to monitor and limit the toxicity effects induced by heavy elements such as mercury. Recently, Li et al. [32] investigated the famous Tibetan medicine called Zuotai, which has been used for treating various diseases for over 1300 years, and is still used in some areas in the Tibetan Plateau. These drugs contain mercury and, counter intuitively, they are not toxic. Zuotai was proven effective in treating inflammations, anthrax, and so on. To understand the mechanisms beyond the Tibetan medicine, Li et al. [8] investigated the chemical speciation as well as the distribution of mercury in the organs of treated animals, by combining the μ-XRF and XANES techniques. By comparing the XANES spectra at Hg L_3-edge for Zuotai and other reference mercury compounds, they recognized $HgCys_2$ and MeHgCys as the mercury compounds present in the organs. By using μ-XRF imaging with a 200 μm spatial resolution, they also revealed that mercury is accumulated in the renal cortex of the kidney. Moreover, the mercury deposition in the kidneys treated with Zuotai is much less than those treated by feeding β-HgS or $HgCl_2$ (Figure 2). The speciation of Hg was revealed with a linear combination fitting with reference to the Hg L_3-edge XANES spectra, and most of the Hg^{2+} ions were bounded to the sulfydryl group, which is less toxic. From this study, we learnt that both the microscopic distribution and the bond nature of Hg ions are important in order to understand the toxicity and the transport mechanisms of Hg in biological systems.

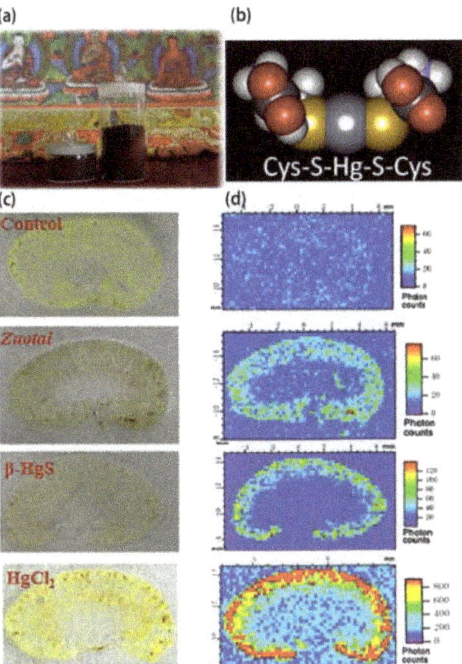

Figure 2. (**a**) Photo of the Tibetan medicine Zuotai; (**b**) molecular scheme of the HgCys$_2$; (**c**) images of slices of kidney from mice untreated and mice treated with Zuotai, β-HgS, or HgCl$_2$, respectively; (**d**) maps of the mercury distribution using μ-X-ray fluorescence (XRF) (figures redrawn from the literature [32] and with courtesy of Cen Li).

The consumption of fish, fish products, and marine mammals is generally considered to be the main global pathway of human exposure to methylmercury (MeHg). Soil is the primary source of MeHg exposure to rice plant tissues. By using the μ-XRF and XANES spectra at the Hg L_3-edge, Meng et al. [33] investigated the distribution and speciation of mercury in three fractions of the rice grains (hull, bran, and white rice), collected from an Hg-contaminated region in China. The majority of the inorganic mercury (IHg) in a rice grain is found in the hull and bran (Figure 3). However, the majority of the toxic species, the methyl mercury (MeHg), is found in edible white rice. This study revealed that the transport of mercury due to anthropic activities is extremely sophisticated. Identifying the atomic coordination of mercury may provide a clue for figuring out a better solution to fix or minimize pollution issues. By employing the high-energy resolution XANES (HR-XANES) spectra [27,34–37] with the fluorescence detection method, Manceau et al. obtained fine details in the near-edge region at the Hg L_3-edge. Combining the theoretical simulations, the bonding nature of the mercury compounds in the environment was investigated with more physical insights [36,37]. A key link between the inorganic Hg input and exposure of humans and wildlife is the net production of methyl-mercury, one of the most dangerous Hg species with regards to toxicity issues. It is mainly produced in the anoxic zones of freshwater, terrestrial, and coastal environments, and in the deep part of the ocean [38]. Climate change is occurring across the world, and in this context, high latitude and altitude regions are particularly fragile, where climate change is advancing faster than in other regions.

The Tibetan Plateau (TP) is the highest and largest highland in the world, with an average elevation of over 4000 m above sea level (a.s.l.). The permafrost body in the TP is the largest permafrost region at low- and mid-latitudes, and accounts for 74.5% of the Northern Hemisphere's mountain permafrost. Presently, limited studies on Hg behaviors (e.g., distribution, variation, etc.) have been performed in high-altitude permafrost regions [39], and, to the best of our knowledge, the influencing factors on Hg behavior in the frozen soils of the permafrost regions have been poorly studied. Moreover, only few studies have reported concentrations and distribution of Hg in the surface soils of the TP. Hg concentrations are typically extremely low, and attempts to measure the Hg K characteristic lines are needed. Although challenging, this approach is strategic to probe the Hg amount in soils and to understand how chemical mechanisms may occur in different regions. Moreover, it is also important to investigate how the different Hg species are transported, transformed, and accumulated at local and global scale.

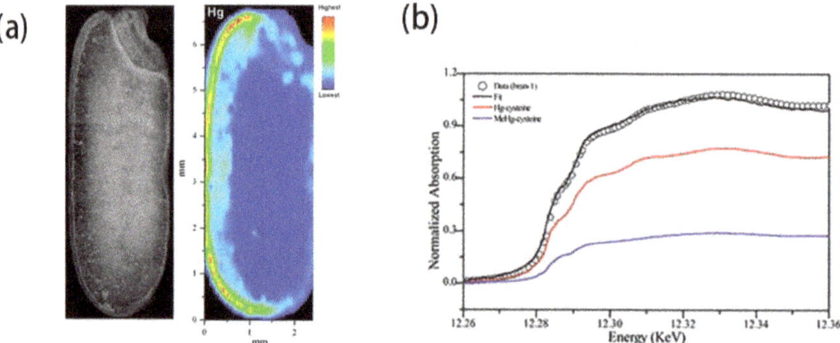

Figure 3. (**a**) Comparison of the optical image with the μ-XRF image and (**b**) comparison of the X-ray absorption near-edge spectroscopy (XANES spectra) of a grain of rice, as well as the reference standards (figures redrawn from the literature [33]). Reprinted with permission from [33]. Copyright {2014} American Chemical Society.

3.2. Spectroscopy and Incineration of Waste

The dust and aerosols originate from the solid particles and/or liquid droplets emitted, as well as the secondary particles that are formed via chemical reactions in the atmosphere. Municipal solid waste (MSW) refers to the trash or garbage from homes, schools, and hospitals, as well as products such as packaging, furniture, clothing, bottles, newspapers, appliances, batteries, and so on. Environmental pollution can be associated with MSW.

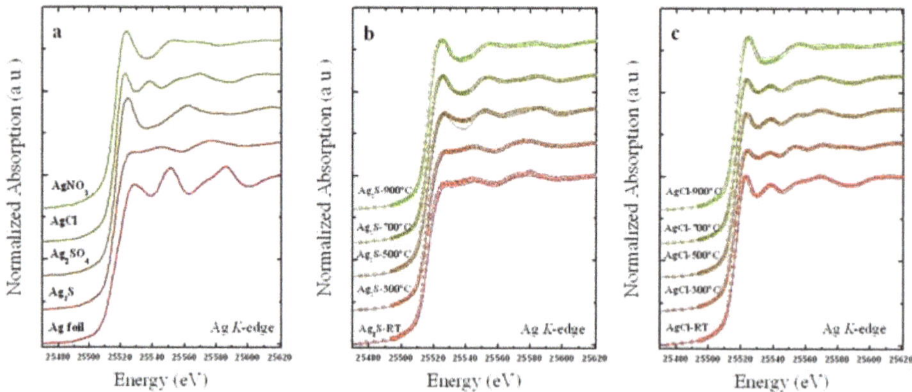

Figure 4. The linear combination fit of the temperature dependent XANES spectra at the Ag *K*-edge: (**a**) reference spectra of reference compounds; (**b**) Ag$_2$S spiked sludge; and (**c**) AgCl spiked simulated municipal solid waste (SMSW). See the text for more details (figure redrawn from the literature [41]).

For instance, burning waste coal can generate hazardous matter, with serious health risks for human beings [40]. It is essential to retrace the origin of these pollutants and to figure out an approach to reduce the amount of hazardous contents. To identify the chemical reactions occurring in the treatment of industrial waste, Yin et al. [41] investigated the chemical transformation of AgCl and Ag_2S in silver nanoparticles under sunlight or incineration. By using the linear combination fitting of the Ag K-edge XANES spectra of Ag_2S spiked sludge and AgCl spiked simulated municipal solid waste (SMSW) incinerated at various temperatures and with certain protocols, it was pointed out that chemical transformations of Ag_2S and AgCl into elemental silver could be a source of Ag nanoparticles (See Figure 4).

Municipal solid waste incineration (MSWI) is a widespread method to treat MSW, and fly ash can be produced in this process. At variance, the secondary fly ash (SFA) is due to a secondary chemical reaction occurring with condensation after vaporization, during the flue gas cooling procedure. The latter is a reusable material suitable for the construction industry to produce bricks. Hence, it is fundamental to investigate the speciation of toxic heavy metals using XAFS. In the SFA samples, Tian et al. [42] revealed that lead mainly existed as $PbCl_2$ and PbS, while PbO and $PbCl_2$ can be identified in MSWI fly ash using a linear combination fitting of the XANES spectra. By using the EXAFS fitting, they also reconstructed the geometrical structure around lead. This study underlines how the structural information is important for optimizing the recycling process of municipal solid waste.

3.3. Applications of Tomography in Environmental Research

Three-dimensional imaging using X-ray micro-computed tomography enables the visualization and quantification of the brine network morphology and variability [43,44]. Duplicate scans have been performed using X-ray energies above and below the absorption edge of the interested elements, using a synchrotron radiation source, which provides monochromatic radiation and a highly tunable beam at variable X-ray energy. Synchrotron radiation X-ray tomography based on absorption contrast enables a clear discrimination of air, ice, and solid salts, if the contained elements have a drastic mass difference. For instance, the internal structure of sea ice samples, shown in Figure 5, can be observed with a strong contrast [45], as follows: air (dark), ice (grey), and salt crystals or brine (white). The field of view was about 12 mm, while the voxel size was 5.6 μm [45]. With the improved algorithms and the advances in X-ray tomography, one can obtain a qualitative estimation of the pore size, and so on, which could be useful to estimate the motion of the glacier [46]. Furthermore, the tomography method can be successfully used to study clathrate gas hydrates [45,47–49], ice and clathrate [50], plant roots and soils [51], aerosols [52], and so on.

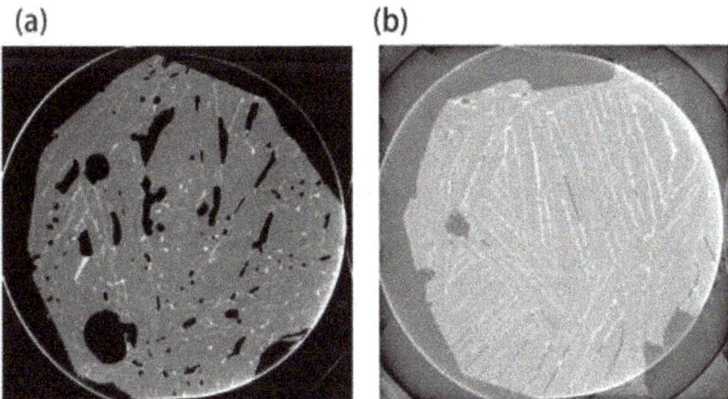

Figure 5. Images of a horizontal slice of a sea ice sample: (**a**) as grown and (**b**) rapidly cooled, as measured by SR tomography at about $-30°$ C (figures redrawn from the literature [45]).

4. Perspectives in Applications of SR Techniques in Cryospheric Sciences

In many cryospheric researches, even the sampling and its preservation are challenging. Samples to characterize and understand the atmosphere, the climate evolution, and the status of the Earth's ecosystem are rare and extremely precious. Actually, airborne particles and dust are generated and transported over time by natural mechanisms and by human related activities. During transport, the elements may experience different thermodynamical conditions, and chemical reactions may occur, affecting the speciation of many elements. To recognize the source as well as the evolution pathways, synchrotron radiation techniques represent a powerful and unique tool. Although a limited number of publications [5–9] in this field have used synchrotron radiation techniques, efforts to expand the number of applications will certainly be beneficial to trigger new researches and to expand the community working in cryospheric science. Indeed, ice and snow are key elements of the cryosphere; ice is central to climate, geology, and life. Understanding its behavior on Earth is essential for predicting the future of the planet and unraveling the emergence of life in the universe [53–56]. Although challenging, the following research opportunities using synchrotron radiation techniques are extremely promising:

(1) trace elemental analysis in deep ice core. By applying SR-XRF, it is possible to obtain quantitative information about the elemental concentration without damaging the sample. However, for ultra-trace analysis, it is important to the protect samples under investigation from external contaminations. Specially designed sample chambers installed at end-stations are required. To reach ultra-low detection limits with the highest S/N ratio, the highest flux at the sample position and the best detectors are necessary;

(2) the speciation of metals, as illustrated in the previous examples, is possible through performing XANES experiments. However, it is necessary to concentrate the investigated elements, and sample preparation is a key issue [57,58]. This is a particularly important to investigate the inorganic fraction contained in the deep ice core [59]. In this case, it is necessary to melt from several centimeters up to meters of ice core to reach the detection limit. Moreover, a spectral database should be established with as many standards as possible to identify the chemical reactions that might occur. A mimicking in situ XANES measurement under different sample conditions (gas, temperature, etc.) would also help to understand the reaction pathways of the different elements induced by climatic changes or anthropogenic activities;

(3) establish the largest possible dataset by measuring samples from the entire ecosystem extending over time and regions. Statistics will be helpful to draw more reliable conclusions. Hence, in order to describe climatic changes, one needs to sample the entire ecosystem, including, but not limited to, ice, snow, aerosols, dust, soils, solid state waste, and so on.

The introduction of new SR techniques and overcoming technical issues are always possible. We need to focus on the emerging techniques so as to investigate at the best overall physical–chemical processes occurring in the cryosphere and in its components (air, water, snow, ice, and permafrost). There are many difficulties that have limited experimental research in this field, but they can be overcome with a synergic effort among scientists and researchers working in these large infrastructures. As shown in Table 1, there are already some beamlines that can offer the XRF, XAFS, XRD, and tomography at current synchrotron radiation facilities in Beijing, Shanghai, and Hefei. The construction of new beamlines in an SSRF upgrade and in planned HEPS would expand the capabilities in many aspects, such as the energy/spatial resolution and integration of different techniques. The community of cryospheric science needs to know the opportunities and contribute.

Table 1. Techniques available at Chinese Synchrotron Facilities * in the mainland of China.

Synchrotron Radiation Facility	Beamline	Availability	Energy (keV)	Focal Spot (V × H µm)	Technique
Beijing Synchrotron Radiation Facility (BSRF)	4W1B	Operation	5–20	20 × 20	µ-XRF
	1W1B	Operation	4–23	900 × 300	XAFS
	1W2B	Operation	5–20	1000 × 600	XAFS
	4B7A	Operation	2.1–5.7	3000 × 1000	XAFS, calibration
	4B9A	Operation	4–15	2000 × 1000	XRD, XAFS,
	4W1A	Operation	6–22	20000–10,000	CT
			5–12	15 × 15	Nano-CT
Shanghai Synchrotron Radiation Facility (SSRF)	BL15U1	Operation	5–20	1.6 × 1.8 (variable) using KB 0.15 × 0.15 using zone plate	µ-XRF µ-XAFS µ-XRD
	BL14W1	Operation	4.5–35	300 × 300	XAFS
	BL14B1	Operation	4–22	400 × 400	XRD
	BL13W1	Operation	8–72.5	45,000 × 5000	CT
Shanghai Synchrotron Radiation Facility (SSRF) Upgrade	E-line	Construction	1.3–10	80 × 200	XPS, XAFS
	D-line	Construction	5–25 (X-ray)	NA	XAFS, FTIR
			$10–10^4$ cm^{-1} (IR)	100 × 100	
	Tender beamline	Construction	2.1~16	5 × 1.5	XAFS
	General spectroscopy	Construction	5–30	500 × 100	XAFS
	Nanobeamline	Construction	5–25	0.01 × 0.01	
National Synchrotron Radiation Laboratory (NSRL)	BL01B	Operation	15–4000 cm^{-1}	NA	FTIR
High Energy Photon Source (HEPS)	BD	Planning	2.1–7.8	400 × 400	XAFS
	BE	Planning	5–15	NA	Tomography
	B2	Planning	5–25	0.009 × 0.009	n-XRF/n-XRD
	B5	Planning	5–25	10 × 10	XRS, NRS, RIXS
	B8	Planning	4.8–45	NA	QXAFS

* BSRF and HEPS are both in Beijing, SSRF in Shanghai, and NSRL in Hefei; * Other sources have been proposed in China, but at present no precise data are available.

5. Relevance of Cryospheric Sciences to Synchrotron Radiation in China

Synchrotron radiation techniques may offer several opportunities in cryosphere and environmental sciences, because of their high resolution and reliability. Chinese scientists first used synchrotron radiation to analyze algae, lichen, bryophyte, and Adelie penguins from Antarctica [60–62]. These early works promoted polar environmental researches at the micro level. The X-ray fluorescence method is extremely powerful to investigate the role of heavy metals on polar species. This method has been also applied in an Arctic aerosol research [63]. Samples of aerosols from the marine boundary layer of the Arctic Ocean were collected aboard the R/V Xuelong on the Second Chinese Arctic Research Expedition (July–September 2003), and the chemical composition of these particles was determined using SR-XRF.

Mercury is a global pollutant that affects human and ecosystem health. In the last two decades, anthropogenic Hg emissions declined in Europe, but significantly increased in Asia [64,65]. This represents a relevant issue for the Arctic region. Many studies also discussed the toxicity of Hg, which is released in the food chain as glaciers retreat. Because high amounts of Br/Cl and O_3 species will react with Hg from atmosphere, the increasing of Hg at the snow surface occurs when atmospheric mercury depletion events occur. Many data have pointed out that snow pack and melt water are the major reservoirs of atmospherically deposited mercury [66,67]. The inductively coupled plasma mass spectrometry (ICP-MS) and general mass spectrometry methods are powerful, but return limited information because they may only measure the total amount of Hg in ice cores. Moreover, this technique does not recognize elemental Hg from $MeHg^+$ and Hg^{2+} [68]. Information on other speciation mechanisms is also lacking. Moreover, the intensity of the solar radiation is very different in Arctic compared with other regions like the Tibet area. By contrast to the polar day in Arctic, a much stronger solar irradiation occurs at the high altitude of the Tibet Plateau. SR studies may allow for a more precise characterization of Hg speciation, pointing out the different reduction and oxidation conditions. Presently, the molecular mechanisms (e.g., molecular Cl and B) beyond these processes are largely unknown [69,70], and reliable knowledge of the chemical reactions in ice and snow is lacking [64]. Actually, the mercury concentration in these samples is extremely low and hard to be detected by a conventional setup. Nevertheless, the improvement of the source brilliance and multi-element detectors will certainly enable the detection of Hg down to the *ng/g* level or lower, to reach the expected values of the natural concentration. Moreover, it is necessary to establish protocols for sample filtering and concentrating, so as to perform XANES spectroscopy experiments on the same samples. These protocols are essential in order to minimize the possible contamination of these precious samples during the experiments. Hence, special sample handling systems should be implemented at dedicated beamline end-stations. Within this framework and the synergic effort among researchers and synchrotron radiation scientists, this interdisciplinary community will be able to investigate more and more samples in the existing as well as in future Chinese light sources.

6. Conclusions

In this contribution, we reviewed some applications of XRF and XAFS techniques in environmental and cryospheric sciences. The trace-element analysis and/or the mapping are now possible using synchrotron radiation. XRF allows for identify the origin and distribution of many elements of interest, ranging from low (Na, Mg, etc.) to high Z elements. Spectroscopic methods such as the XAFS technique offer a new perspective on the speciation of elements, a condition that may affect the biological functioning of many systems, and the environmental release or particulate matters. Furthermore, high-energy resolution spectroscopy may reveal detailed spectral features that could be used to better investigate and recognize the nature of many atomic bonds. Finally, new powerful imaging methods, such as X-ray tomography, could provide three-dimensional views of the internal structure of many systems, including ice and clathrate systems.

Presently, the applications of synchrotron radiation techniques in cryospheric sciences are still limited, although some recent publications [5–9] and workshops [59,71] demonstrated that the cooperation among scientists working in cryospheric sciences and synchrotron radiation researchers is possible and extremely useful [65]. An enhanced collaboration is foreseen and efforts are necessary to make this highly interdisciplinary research area mature in order to advance the frontiers of this challenging discipline. The role of Chinese facilities and scientists will be certainly important in the next years [72].

Author Contributions: Z.D., A.M., W.X., and C.X. designed the study. Z.D., W.X., A.M., S.L., and Y.Z. carried out the synchrotron radiation experiments at BSRF, ESRF, Diamond, and ALBA over the years. W.X., A.M., S.L., and Z.D. wrote the manuscript with contributions from all of the authors.

Funding: This work was supported by the National Natural Science Foundation of China (grant No. U1532128, no. 41425003, and no. 41701071)

Acknowledgments: W.X. acknowledges the financial support and the hospitality of LNF under the framework of IHEP and INFN collaboration. W.X is grateful to Cen Li, Bo Meng, Yongguang Yin, and Jiating Zhao for sharing their ideas and insightful discussions. We strongly acknowledge the staff of the Italian CRG LISA for their support at ESRF within the experiment 08-01-1031 on Beamline BM08, and for many fruitful discussions.

Conflicts of Interest: No conflicts of interest.

References

1. Qin, D.; Ding, Y.; Xiao, C.; Kang, S.; Ren, J.; Yang, J.; Zhang, S. Cryospheric science: Research framework and disciplinary system. *Natl. Sci. Rev.* **2018**, *5*, 255–268. [CrossRef]

2. Zhang, X.; Li, H.; Zhang, Z.; Wu, Q.; Zhang, S. Recent glacier mass balance and area changes from dems and landsat images in upper reach of shule river basin, northeastern edge of tibetan plateau during 2000 to 2015. *Water* **2018**, *10*, 796. [CrossRef]

3. Yin, X.; Kang, S.; de Foy, B.; Ma, Y.; Tong, Y.; Zhang, W.; Wang, X.; Zhang, G.; Zhang, Q. Multi-year monitoring of atmospheric total gaseous mercury at a remote high-altitude site (nam co, 4730ma.S.L.) in the inland tibetan plateau region. *Atmos. Chem. Phys.* **2018**, *18*, 10557–10574. [CrossRef]

4. Jiang, X.; Wang, J.; Qin, Q.; Dong, Y.; Sheng, W.; Cheng, J.; Xu, G.; Hu, T.; Deng, H.; Chen, F.; et al. The chinese high-energy photon source and its r&d project. *Synchrotron Radiat. News* **2014**, *27*, 27–31.

5. Marcelli, A.; Hampai, D.; Cibin, G.; Maggi, V. Local vs global climate change: Investigation of dust from deep ice cores. *Spectrosc. Eur.* **2012**, *24*, 12–17.

6. Marcelli, A.; Cibin, G.; Hampai, D.; Maggi, V. Mineralogical characterization of the inorganic component from deep ice core samples: A challenging xanes investigation. *IXAS Res. Rev.* **2012**, *8*. Available online: https://www.ixasportal.net/ixas/index.php?option=com_content&view=article&id=23&Itemid=373# (accessed on 30 September 2018).

7. Cibin, G.; Marcelli, A.; Maggi, V.; Sala, M.; Marino, F.; Delmonte, B.; Albani, S.; Pignotti, S. First combined total reflection X-ray fluorescence and grazing incidence X-ray absorption spectroscopy characterization of aeolian dust archived in antarctica and alpine deep ice cores. *Spectrochim. Acta Part B At. Spectrosc.* **2008**, *63*, 1503–1510. [CrossRef]

8. Marcelli, A.; Hampai, D.; Giannone, F.; Sala, M.; Maggi, V.; Marino, F.; Pignotti, S.; Cibin, G. Xrf-xanes characterization of deep ice core insoluble dust. *J. Anal. At. Spectrom.* **2012**, *27*, 33–37. [CrossRef]

9. Ventura, G.D.; Marcelli, A.; Bellatreccia, F. Sr-ftir microscopy and ftir imaging in the earth sciences. *Rev. Min. Geochem.* **2014**, *78*, 447–479. [CrossRef]

10. Obbard, R.W.; Lieb-Lappen, R.M.; Nordick, K.V.; Golden, E.J.; Leonard, J.R.; Lanzirotti, A.; Newville, M.G. Synchrotron X-ray fluorescence spectroscopy of salts in natural sea ice: Sxrf of salts in natural sea ice. *Earth Space Sci.* **2016**, *3*, 463–479. [CrossRef]

11. Zhang, L.L.; Yan, S.; Jiang, S.; Yang, K.; Wang, H.; He, S.; Liang, D.X.; Zhang, L.; He, Y.; Lan, X.Y.; et al. Hard X-ray micro-focusing beamline at ssrf. *Nucl. Sci. Tech.* **2015**, *26*, 060101–060107.

12. Solé, V.A.; Papillon, E.; Cotte, M.; Walter, P.; Susini, J. A multiplatform code for the analysis of energy-dispersive X-ray fluorescence spectra. *Spectrochim. Acta Part B At. Spectrosc.* **2007**, *62*, 63–68. [CrossRef]

13. Hartree, D.R.; Kronig, R.D.; Petersen, H. A theoretical calculation of the fine structure for the k-absorption band of ge in gecl4. *Physica* **1934**, *1*, 895–924. [CrossRef]

14. Rehr, J.J.; Albers, R.C. Theoretical approaches to X-ray absorption fine structure. *Rev. Mod. Phys.* **2000**, *72*, 621–654. [CrossRef]

15. Dill, D.; Dehmer, J.L. Electron-molecule scattering and molecular photoionization using the multiple-scattering method. *J. Chem. Phys.* **1974**, *61*, 692–699. [CrossRef]

16. Balzarotti, A.; Bianconi, A.; Burattini, E.; Grandolfo, M.; Habel, R.; Piacentini, M. Core transitions from the al 2p level in amorphous and crystalline Al_2O_3. *Phys. Status Solidi (B)* **1974**, *63*, 77–87. [CrossRef]

17. Bianconi, A.; Petersen, H.; Brown, F.C.; Bachrach, R.Z. K-shell photoabsorption spectra of N_2 and N_2O using synchrotron radiation. *Phys. Rev. A* **1978**, *17*, 1907–1911. [CrossRef]

18. Bianconi, A.; Doniach, S.; Lublin, D. X-ray ca k edge of calcium adenosine triphosphate system and of simple ca compunds. *Chem. Phys. Lett.* **1978**, *59*, 121–124. [CrossRef]

19. Bianconi, A. Core excitons and inner well resonances in surface soft X-ray absorption (ssxa) spectra. *Surface Sci.* **1979**, *89*, 41–50. [CrossRef]

20. Belli, M.; Scafati, A.; Bianconi, A.; Mobilio, S.; Palladino, L.; Reale, A.; Burattini, E. X-ray absorption near edge structures (xanes) in simple and complex mn compounds. *Solid State Commun.* **1980**, *35*, 355–361. [CrossRef]

21. Benfatto, M.; Natoli, C.R.; Bianconi, A.; Garcia, J.; Marcelli, A.; Fanfoni, M.; Davoli, I. Multiple-scattering regime and higher-order correlations in X-ray-absorption spectra of liquid solutions. *Phys. Rev. B* **1986**, *34*, 5774–5781. [CrossRef]

22. Longa, S.D.; Arcovito, A.; Girasole, M.; Hazemann, J.L.; Benfatto, M. Quantitative analysis of X-ray absorption near edge structure data by a full multiple scattering procedure: The fe-co geometry in photolyzed carbonmonoxy-myoglobin single crystal. *Phys. Rev. Lett.* **2001**, *87*, 155501–155504. [CrossRef] [PubMed]

23. Ravel, B.; Newville, M. Athena, artemis, hephaestus: Data analysis for X-ray absorption spectroscopy using ifeffit. *J. Synchrotron Radiat.* **2005**, *12*, 537–541. [CrossRef] [PubMed]

24. Ankudinov, A.L.; Ravel, B.; Rehr, J.J.; Conradson, S.D. Real-space multiple-scattering calculation and interpretation of X-ray-absorption near-edge structure. *Phys. Rev. B* **1998**, *58*, 7565. [CrossRef]

25. Joly, Y. X-ray absorption near-edge structure calculations beyond the muffin-tin approximation. *Phys. Rev. B* **2001**, *63*. [CrossRef]

26. Benfatto, M.; Della Longa, S. Geometrical fitting of experimental xanes spectra by a full multiple-scattering procedure. *J. Synchrotron Radiat.* **2001**, *8*, 1087–1094. [CrossRef] [PubMed]

27. Manceau, A.; Lemouchi, C.; Enescu, M.; Gaillot, A.-C.; Lanson, M.; Magnin, V.; Glatzel, P.; Poulin, B.A.; Ryan, J.N.; Aiken, G.R.; et al. Formation of mercury sulfide from hg(ii)–thiolate complexes in natural organic matter. *Environ. Sci. Technol.* **2015**, *49*, 9787–9796. [CrossRef] [PubMed]

28. Török, S.; Faigel, G.; Jones, K.W.; Rivers, M.L.; Sutton, S.R.; Bajt, S. Chemical characterization of environmental particulate matter using synchrotron radiation. *X-ray Spectrom.* **1994**, *23*, 3–6. [CrossRef]

29. Momose, A.; Takeda, T.; Itai, Y.; Hirano, K. Phase-contrast X-ray computed tomography for observing biological soft tissues. *Nat. Med.* **1996**, *2*, 473–475. [CrossRef] [PubMed]

30. Munro, P.R.T.; Ignatyev, K.; Speller, R.D.; Olivo, A. Phase and absorption retrieval using incoherent X-ray sources. *Proc. Natl. Acad. Sci. USA* **2012**, *109*, 13922. [CrossRef] [PubMed]

31. Sen, I.S.; Peucker-Ehrenbrink, B. Anthropogenic disturbance of element cycles at the earth's surface. *Environ. Sci. Technol.* **2012**, *46*, 8601–8609. [CrossRef] [PubMed]

32. Li, C.; Xu, W.; Chu, S.; Zheng, Z.; Xiao, Y.; Li, L.; Bi, H.; Wei, L. The chemical speciation, spatial distribution and toxicity of mercury from tibetan medicine zuotai, β-hgs and hgcl2 in mouse kidney. *J. Trace Elem. Med. Biol.* **2018**, *45*, 104–113. [CrossRef] [PubMed]

33. Meng, B.; Feng, X.; Qiu, G.; Anderson, C.W.N.; Wang, J.; Zhao, L. Localization and speciation of mercury in brown rice with implications for pan-asian public health. *Environ. Sci. Technol.* **2014**, *48*, 7974–7981. [CrossRef] [PubMed]

34. Manceau, A. Comment on "direct observation of tetrahedrally coordinated fe(iii) in ferrihydrite". *Environ. Sci. Technol.* **2012**, *46*, 6882–6884. [CrossRef] [PubMed]

35. Manceau, A.; Enescu, M.; Simionovici, A.; Lanson, M.; Gonzalez-Rey, M.; Rovezzi, M.; Tucoulou, R.; Glatzel, P.; Nagy, K.L.; Bourdineaud, J.-P. Chemical forms of mercury in human hair reveal sources of exposure. *Environ. Sci. Technol.* **2016**, *50*, 10721–10729. [CrossRef] [PubMed]

36. Manceau, A.; Lemouchi, C.; Rovezzi, M.; Lanson, M.; Glatzel, P.; Nagy, K.L.; Gautier-Luneau, I.; Joly, Y.; Enescu, M. Structure, bonding, and stability of mercury complexes with thiolate and thioether ligands from high-resolution xanes spectroscopy and first-principles calculations. *Inorg. Chem.* **2015**, *54*, 11776–11791. [CrossRef] [PubMed]

37. Manceau, A.; Wang, J.; Rovezzi, M.; Glatzel, P.; Feng, X. Biogenesis of mercury–sulfur nanoparticles in plant leaves from atmospheric gaseous mercury. *Environ. Sci. Technol.* **2018**, *52*, 3935–3948. [CrossRef] [PubMed]

38. Driscoll, C.T.; Mason, R.P.; Chan, H.M.; Jacob, D.J.; Pirrone, N. Mercury as a global pollutant: Sources, pathways, and effects. *Environ. Sci. Technol.* **2013**, *47*, 4967–4983. [CrossRef] [PubMed]

39. Sun, S.; Kang, S.; Huang, J.; Chen, S.; Zhang, Q.; Guo, J.; Liu, W.; Neupane, B.; Qin, D. Distribution and variation of mercury in frozen soils of a high-altitude permafrost region on the northeastern margin of the tibetan plateau. *Environ. Sci. Pollut. Res.* **2017**, *24*, 15078–15088. [CrossRef] [PubMed]

40. Whiteside, M.; Herndon, J. Coal fly ash aerosol: Risk factor for lung cancer. *J. Adv. Med. Med. Res.* **2018**, *25*, 1–10. [CrossRef]

41. Yin, Y.; Xu, W.; Tan, Z.; Li, Y.; Wang, W.; Guo, X.; Yu, S.; Liu, J.; Jiang, G. Photo- and thermo-chemical transformation of agcl and ag2s in environmental matrices and its implication. *Environ. Pollut.* **2017**, *220*, 955–962. [CrossRef] [PubMed]

42. Tian, S.; Zhu, Y.; Meng, B.; Guan, J.; Nie, Z.; Die, Q.; Xu, W.; Yu, M.; Huang, Q. Chemical speciation of lead in secondary fly ash using X-ray absorption spectroscopy. *Chemosphere* **2018**, *197*, 362–366. [CrossRef] [PubMed]

43. Lieb-Lappen, R.M.; Golden, E.J.; Obbard, R.W. Metrics for interpreting the microstructure of sea ice using X-ray micro-computed tomography. *Cold Reg. Sci. Technol.* **2017**, *138*, 24–35. [CrossRef]

44. Lieblappen, R.M.; Kumar, D.D.; Pauls, S.D.; Obbard, R.W. A network model for characterizing brine channels in sea ice. *Cryosphere* **2018**, *12*, 1013–1026. [CrossRef]

45. Maus, S.; Huthwelker, T.; Enzmann, F.; M Miedaner, M.; Stampanoni, M.; Marone, F.; Hutterli, M.; Hintermüller, C.; Kersten, M. Synchrotron-based X-ray micro-tomography: Insights into sea ice microstructure. In Proceedings of the Sixth Workshop on Baltic Sea Ice Climate, Lammi Biological Station, Finland, 2008; 2009; Volume 61, pp. 28–45.

46. Faria, S.H.; Weikusat, I.; Azuma, N. The microstructure of polar ice. Part ii: State of the art. *J. Struct. Geol.* **2014**, *61*, 21–49. [CrossRef]

47. Murshed, M.M.; Klapp, S.A.; Enzmann, F.; Szeder, T.; Huthwelker, T.; Stampanoni, M.; Marone, F.; Hintermüller, C.; Bohrmann, G.; Kuhs, W.F.; et al. Natural gas hydrate investigations by synchrotron radiation X-ray cryo-tomographic microscopy (srxctm). *Geophys. Res. Lett.* **2008**, *35*. [CrossRef]

48. Takeya, S.; Honda, K.; Gotoh, Y.; Yoneyama, A.; Ueda, K.; Miyamoto, A.; Hondoh, T.; Hori, A.; Sun, D.; Ohmura, R.; et al. Diffraction-enhanced X-ray imaging under low-temperature conditions: Non-destructive observations of clathrate gas hydrates. *J. Synchrotron Radiat.* **2012**, *19*, 1038–1042. [CrossRef] [PubMed]

49. Yang, L.; Zhao, J.; Liu, W.; Li, Y.; Yang, M.; Song, Y. Microstructure observations of natural gas hydrate occurrence in porous media using microfocus X-ray computed tomography. *Energy Fuels* **2015**, *29*, 4835–4841. [CrossRef]

50. Arzbacher, S.; Petrasch, J.; Ostermann, A.; Loerting, T. Micro-tomographic investigation of ice and clathrate formation and decomposition under thermodynamic monitoring. *Materials* **2016**, *9*, 668. [CrossRef] [PubMed]

51. Koebernick, N.; Daly, K.R.; Keyes, S.D.; George, T.S.; Brown, L.K.; Raffan, A.; Cooper, L.J.; Naveed, M.; Bengough, A.G.; Sinclair, I.; et al. High-resolution synchrotron imaging shows that root hairs influence rhizosphere soil structure formation. *New Phytol.* **2017**, *216*, 124–135. [CrossRef] [PubMed]

52. Porra, L.; Dégrugilliers, L.; Broche, L.; Albu, G.; Strengell, S.; Suhonen, H.; Fodor, G.H.; Peták, F.; Suortti, P.; Habre, W.; et al. Quantitative imaging of regional aerosol deposition, lung ventilation and morphology by synchrotron radiation ct. *Sci. Rep.* **2018**, *8*, 3519. [CrossRef] [PubMed]

53. Bartels-Rausch, T. Ten things we need to know about ice and snow. *Nature* **2013**, *494*, 27–29. [CrossRef] [PubMed]

54. Bartels-Rausch, T.; Bergeron, V.; Cartwright, J.H.E.; Escribano, R.; Finney, J.L.; Grothe, H.; Gutiérrez, P.J.; Haapala, J.; Kuhs, W.F.; Pettersson, J.B.C.; et al. Ice structures, patterns, and processes: A view across the icefields. *Rev. Mod. Phys.* **2012**, *84*, 885–944. [CrossRef]

55. Nakamura, T.; Noguchi, T.; Tsuchiyama, A.; Ushikubo, T.; Kita, N.T.; Valley, J.W.; Zolensky, M.E.; Kakazu, Y.; Sakamoto, K.; Mashio, E.; et al. Chondrulelike objects in short-period comet 81p/wild 2. *Science* **2008**, *321*, 1664. [CrossRef] [PubMed]

56. Fitzner, M.; Sosso, G.C.; Cox, S.J.; Michaelides, A. The many faces of heterogeneous ice nucleation: Interplay between surface morphology and hydrophobicity. *J. Am. Chem. Soc.* **2015**, *137*, 13658–13669. [CrossRef] [PubMed]

57. Macis, S.; Cibin, G.; Maggi, V.; Baccolo, G.; Hampai, D.; Delmonte, B.; D'Elia, A.; Marcelli, A. Microdrop deposition technique: Preparation and characterization of diluted suspended particulate samples. *Condens. Matter* **2018**, *3*, 21. [CrossRef]

58. D'Elia, A.; Cibin, G.; Robbins, P.E.; Maggi, V.; Marcelli, A. Design and characterization of a mapping device optimized to collect xrd patterns from highly inhomogeneous and low density powder samples. *Nucl. Instrum. Methods Phys. Res. Sec. B Beam Interact. Mater. Atoms* **2017**, *411*, 22–28. [CrossRef]

59. From glacier to climate—Euro-Asian perspectives in cryospheric sciences. In Proceedings of the Bilateral Chinese/Italian Workshop, Beijing, China, 9–10 July 2012.

60. Shen, X.; Sun, L.; Zhang, L.; Yin, X.; Kang, S.; Wu, Z.; Ju, X.; Huang, Y. Analysis on the 6 species of alagae and lichen by sr-xrf in the fileds peninsula of antarctica. *Chin. J. Pol. Res.* **2001**, *13*, 187–194.

61. Shen, X.; Sun, L.; Yin, X.; Zhang, L.; Kang, S.; Wu, Z.; Huang, Y.; Ju, X. X-ray fluorescent analysis of the 6 species of bryophyte in the king george iland, antarctica. *Chin. J. Pol. Res.* **2001**, *13*, 50–56.

62. Xie, Z.; Sun, L.; Long, N.; Li, Z.; Kang, S.; Wu, Z.; Huang, Y.; Xin, J. Analysis of the distribution of chemical elements in adelie penguin bone using synchrotron radiation X-ray fluorescence. *Pol. Biol.* **2003**, *26*, 171–177.

63. Xie, Z.; Sun, L.; Blum, J.D.; Huang, Y.; He, W. Summertime aerosol chemical components in the marine boundary layer of the arctic ocean. *J. Geophys. Rese. Atmos.* **2006**, *111*. [CrossRef]

64. Liao, J.; Huey, L.G.; Liu, Z.; Tanner, D.J.; Cantrell, C.A.; Orlando, J.J.; Flocke, F.M.; Shepson, P.B.; Weinheimer, A.J.; Hall, S.R.; et al. High levels of molecular chlorine in the arctic atmosphere. *Nat. Geosci.* **2014**, *7*, 91–94. [CrossRef]

65. Shi, Z.; Krom, M.D.; Jickells, T.D.; Bonneville, S.; Carslaw, K.S.; Mihalopoulos, N.; Baker, A.R.; Benning, L.G. Impacts on iron solubility in the mineral dust by processes in the source region and the atmosphere: A review. *Aeolian Res.* **2012**, *5*, 21–42. [CrossRef]

66. Domine, F.; Cincinelli, A.; Bonnaud, E.; Martellini, T.; Picaud, S. Adsorption of phenanthrene on natural snow. *Environ. Sci. Technol.* **2007**, *41*, 6033–6038. [CrossRef] [PubMed]

67. Schroeder, W.H.; Anlauf, K.G.; Barrie, L.A.; Lu, J.Y.; Steffen, A.; Schneeberger, D.R.; Berg, T. Arctic springtime depletion of mercury. *Nature* **1998**, *394*, 331. [CrossRef]

68. Jitaru, P.; Gabrielli, P.; Marteel, A.; Plane, J.M.C.; Planchon, F.A.M.; Gauchard, P.-A.; Ferrari, C.P.; Boutron, C.F.; Adams, F.C.; Hong, S.; et al. Atmospheric depletion of mercury over antarctica during glacial periods. *Nat. Geosci.* **2009**, *2*, 505–508. [CrossRef]

69. Jeong, D.; Kim, K.; Choi, W. Accelerated dissolution of iron oxides in ice. *Atmos. Chem. Phys.* **2012**, *12*, 11125–11133. [CrossRef]

70. Pratt, K.A.; DeMott, P.J.; French, J.R.; Wang, Z.; Westphal, D.L.; Heymsfield, A.J.; Twohy, C.H.; Prenni, A.J.; Prather, K.A. In situ detection of biological particles in cloud ice-crystals. *Nat. Geosci.* **2009**, *2*, 398–401. [CrossRef]

71. Marcelli, A.; Maggi, V. *Aerosols in Snow and Ice. Markers of Environmental Pollution and Climatic Changes: European and Asian Perspectives*; Superstripes Press: Rome, Italy, 2017.

72. Marcelli, A. The large research infrastructures of the people's republic of china: An investment for science and technology. *Phys. Status Solidi (B)* **2013**, *251*, 1158–1168. [CrossRef]

Article

Challenging X-ray Fluorescence Applications for Environmental Studies at XLab Frascati

Giorgio Cappuccio [1], Giannantonio Cibin [2], Sultan B. Dabagov [1,3,4], Alfredo Di Filippo [5], Gianluca Piovesan [5], Dariush Hampai [1,*], Valter Maggi [6,7] and Augusto Marcelli [1,8]

[1] INFN-LNF, XLab Frascati, Via E. Fermi 40, I-00044 Rome, Italy; cappuccio.giorgio@gmail.com (G.C.); sultan.dabagov@lnf.infn.it (S.B.D.); augusto.marcelli@lnf.infn.it (A.M.)
[2] Diamond Light Source Ltd., Harwell Science and Innovation Campus, Didcot OX11 0DE, UK; giannantonio.cibin@diamond.ac.uk
[3] RAS P.N. Lebedev Physical Institute, Moscow 119991, Russia
[4] National Research Nuclear University MEPhI, Moscow 115409, Russia
[5] Dendrology Lab, Department of Agriculture and Forestry Science (DAFNE), University Tuscia, 01100 Viterbo, Italy; difilippo@unitus.it (A.D.F.); piovesan@unitus.it (G.P.)
[6] Earth and Environmental Sciences Department, University of Milano-Bicocca, 20126 Milano, Italy; valter.maggi@unimib.it
[7] INFN Section of Milano-Bicocca, 20126 Milano, Italy
[8] RICMASS, Rome International Center for Materials Science Superstripes, Via dei Sabelli 119A, 00185 Roma, Italy
* Correspondence: dariush.hampai@lnf.infn.it; Tel.: +39-06-9403-5248

Received: 31 July 2018; Accepted: 12 October 2018; Published: 18 October 2018

Abstract: In this work, we will report applications of the total external X-ray fluorescence (TXRF) station, a prototype assembled at the XLab Frascati laboratory (XlabF) at the INFN National Laboratories of Frascati (INFN LNF). XlabF has been established as a facility to study, design and develop X-ray optics, in particular, polycapillary lenses, as well as to perform X-ray experiments for both elemental analysis and tomography. The combination of low-power conventional sources and policapillary optics allows assembling a prototype that can provide a quasi-parallel intense beam for detailed X-ray spectroscopic analysis of extremely low concentrated samples, down to ng/g. We present elemental analysis results of elements contained in tree rings and of dust stored in deep ice cores. In addition to performing challenging environmental research studies, other experiments aim to characterize novel optics and to evaluate original experimental schemes for X-ray diffraction (XRD), X-ray fluorescence (XRF and TXRF) and X-ray imaging.

Keywords: TXRF; polycapillary optics; low concentration elemental analysis

---

1. Introduction

Due to the large penetration depth of X-rays, a variety of analytical techniques, such as X-ray fluorescence (XRF), X-ray diffraction (XRD) and 3D tomography may be used to study the inner sample structure without any destructive preparation [1–3]. Among these techniques, XRF is a method widely used in many different fields to characterize qualitatively and quantitatively the elemental composition of the samples in a non-destructive way (for example, [4,5] and related).

The continuous development of large facilities, such as Synchrotron Radiation (SR) and Free Electron Laser (FEL), together with the significant improvements in X-ray optics greatly expanded the application X-ray techniques for elemental analysis, spectroscopy and imaging studies. In particular, XRF is now commonly utilized as an *ab initio* technique both for qualitative and quantitative analysis, to investigate the composition of a sample down to very low concentrations [6,7]. However, the growth of applications and of users makes the number of available SR sources insufficient to accommodate the requests and many beamlines devoted to X-ray studies are fully packed. As a consequence SR facilities are not available for routine analysis and for many studies the only alternative approach is the use of powerful X-ray microfocus sources coupled to polycapillary optics (PolyCO) [8]. Indeed, the combination of a polycapillary lens with a fine-focus X-ray tube makes it possible to achieve a focusing spot with diameters less than 50 μm. Moreover, it can provide the intensity radiation flux necessary to perform challenging elemental analysis [9]. One of the most attractive applications of such configuration is the confocal geometry with two conjugated X-ray polycapillary lenses [2,9,10]. This peculiar geometry allows one to control the depth of the XRF signal leading to the possibility of depth profile analysis. Due to the high-intensity irradiation on a limited sample area, both non-destructive μXRF analysis and 2D/3D elemental mapping are now possible. Moreover, in the total external reflection X-ray fluorescence regime (TXRF) [11], the decreased detection limits and the lower noises are directly correlated to the lower measurable concentration. To apply this special technique in a simpler way, an incident parallel beam is required. This can be achieved by replacing the primary polycapillary full-lens with a semi-lens, which transforms the divergent source beam into a quasi-parallel one. As shown below, this technique enables excellent results, sometime comparable with those obtained at SR facilities, even for samples with low elemental concentrations [6,7,12].

Among the most interesting applications of this technique is elemental analysis for environmental research. This is a highly demanding application not only in term of recognition of low Z (Na, Al, Mg, K, Ca, etc.) and high Z elements (Fe, Mn, As, Hg, etc.) but also for the large number of samples investigated, sometimes up to hundreds. In the latter case, the time request (several shifts) is hard to fulfill at synchrotron radiation facilities and sometimes the process requires more than one beamline. Moreover, the number of available SR beamlines suitable for these studies is relatively limited and many of them are oversubscribed. Consequently, the use of SR facilities for proposals highly demanding in time and for routine analysis is not straightforward and environmental and climatic studies are not easily performed at these large facilities. As an alternative, a promising approach is the use of X-ray microfocus sources enhanced by means of polycapillary optics [8]. In this contribution, we will show in Section 3.1 some representative examples of tree rings [13] and deep ice cores [14–16] analyzed with an experimental layout based on a combination of polycapillary lens and fine-focus X-ray tubes that can provide high-intensity radiation fluxes. Operating in a total external reflection regime (TXRF), the reduction of noise allows for the improvement of the detection limits, which is a fundamental aspect to be considered in the analysis of a sample with low elemental concentrations. To apply this special technique in a simpler way, an incident parallel beam is required. This can be achieved by replacing the primary polycapillary full-lens with a semi-lens, which converts the divergent source beam into a quasi-parallel one.

The analysis of tree rings is a powerful historical ecology tool for identifying and quantifying important relationships between tree growth and climatic events [17]. Similar to tree rings, deep ice cores taken from polar and continental glaciers are other fundamental natural archives of paleoclimatic information, which can provide unaltered records of environmental and climatic changes at different time scales dating back to several hundred thousand years. The mineralogical composition as well as the size distribution of particles are mainly related to the selection occurring in the atmosphere during dust transport. Nowadays, aeolian mineral dust in thin sections of the Antarctic deep ice cores is actively studied in order to assess the climate changes in the Late Quaternary in the Southern Hemisphere [14,15].

2. XLab Frascati

Since early 2000, a great effort has been devoted at the LNF-INFN to make available the XLab Frascati (XlabF) [18], an optical laboratory dedicated to the study of X-ray optics, in particular, polycapillary lenses and other exotic optical elements. Presently, two facility stations (RXR and XENA, Table 1) are available to users [18], optimizing and matching different analytical techniques. In particular, the XENA (X-ray experimental station for non-destructive analysis) station, operative since 2004, is equipped with three X-ray Oxford Apogee tubes (W, Mo and Cu anodes), a set of mechanical components and motors for lens alignment and scanning and an optical table with different geometrical layouts. At first it was used for all X-ray analysis. Now the facility is dedicated to imaging, tomography and characterization of X-ray devices, such as novel sources [19], optics [20], diffractive crystals [21], vibrating systems [22], fast processes [23], high-resolution imaging [24] and new detectors [25].

Table 1. XLab frascati facilities.

	XENA	**RXR**
Station	X-ray Elemental station for Non-destructive Analysis	Rainbow X-ray
Analysis	(1) High Resolution Imaging (2) μCT (3) X-ray Optics Characterization (4) Detector Characterization (5) Novel Sources	(1) μXRF (2D and 3D mapping) (2) TXRF
Resolution	(1) <1 μm [24] (2) $<17 \times 17 \times 17 \ \mu m^3$ [23]	(1) $\sim 80 \times 80 \ \mu m^2$ [9] $\sim 80 \times 80 \times 80 \ \mu m^3$ [9] (2) 25 ± 1.25 ng/g concentrations [20]

The RXR (Rainbow X-ray) station is an apparatus optimized for 2D/3D XRF micro-imaging and TXRF [9,10]. It is equipped with two detectors characterized by different energy efficiency, covering the wide spectral range from 800 eV to 25 keV. The station works in the confocal geometry, i.e., the source is coupled to a full-lens and both detectors have dedicated half-lenses to match the emission. Samples can be investigated in air or inside a dedicated vacuum chamber that may extend the working capabilities down to the low energy.

A TXRF prototype spectrometer using a polycapillary semi-lens (the layout and its parameters are respectively in Figure 1 and in Table 2) has also been designed and assembled at XlabF. The optics are characterized according to the protocol developed in our lab and described in [20]. The optimum input focal distance of the lens used has been determined by performing a spatial scan of the lens with respect to the X-ray source, while the transmitted radiation has been monitored with a scintillator. The low divergence of this instrument allows one to detect the radiation peak intensity at ~40 cm from the lens exit.

Figure 1. Total external reflection X-ray fluorescence (TXRF) prototype of the X-ray spectrometer. The vacuum chamber works down 10^{-5} hPa, while the positioning system allows independent translations and rotations along six axes $xyz\theta\phi\chi$.

Table 2. Parameters of the TXRF experimental layout.

2 X-ray Tubes [26]	Oxford Apogee 5000 - MoKα - W bremsstrahlung inelastic radiation source spot <50 μm power: 50 W
Silicon Drift Detector [27]	active area: 30 mm^2 with AP3.7 window
CCD Camera **(FDI 1:1.61)** [28]	pixel resolution: 10.4 × 10.4 μm^2
Polycapillary semi-lens [20]	Focal Distance: 59 mm transmission ~60% residual divergence: 1.4 mrad

In this layout, for measurements at low energies, the sample can be positioned in a vacuum chamber working down to 10^{-5} hPa. Inside the vacuum chamber, the positioning is performed by means of a "hexapode micro-positioner" [29]. Such a device allows the translations along six axes $xyz\theta\phi\chi$ with a resolution of 0.5 μm and 19 μrad, Figure 1.

3. TXRF Results

The main experimental parameters for a TXRF prototype are the mass resolution and the minimum detection limit achievable in low concentrated samples. To determine these figures we analyzed standards for XFR, TXRF (Figure 2) by Axo [30] (Table on Supplementary Materials) and by NIST [9].

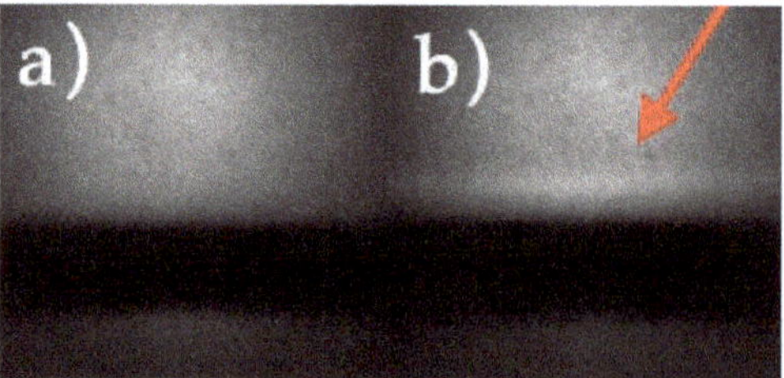

Figure 2. CCD images of the incident beam on AXO sample coated on a silicon wafer): (**a**) out of the total reflection regime and (**b**) in the TXRF regime. The reflected beam is indicated by a red arrow. The black area is due to absorption by the sample.

In Figure 3, the TXRF spectra of two different sources are compared. In this way it is possible to resolve both the elastic (MoKα) and anelastic radiation (W Bremsstrahlung) from the anode. The AXO sample spectrum was provided by the company and collected using a synchrotron radiation source [30].

Neglecting the variation in the spectra of the elastic and the inelastic peaks due to the different sources, the results are in good agreement with the AXO data. Moreover, the energy resolution is improved with respect to the AXO spectrum (e.g., in Figure 3 together with the Pd–L and La–L lines and the splitting of the FeKα and FeKβ).

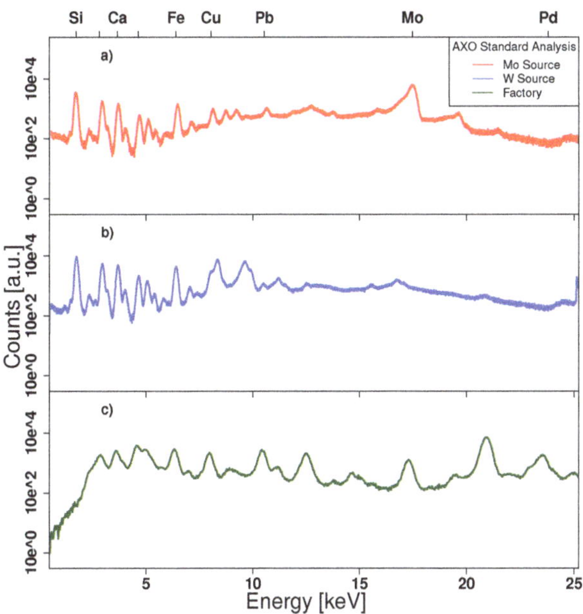

Figure 3. TXRF standard spectra from AXO. Red and blue data are obtained with our TXRF experimental setup, changing only the source anode ((**a**)—MoKα and (**b**)—W Bremsstrahlung respectively), while green (**c**) is the AXO data obtained with synchrotron radiation. Ruling out the elastic and inelastic peaks due to the source, our spectra are in good agreement with AXO data.

3.1. Environmental Analysis

Several natural archives may provide unaltered records of environmental and climatic information at different time scales dating back up to several hundred thousand years: marine sediments [31,32], stalagmite conformations [33], tree rings [13] and deep ice cores [14–16]. The analysis of tree rings is a powerful historical ecology tool because in tree species, inter- and intra-annual responses of radial growth to local climatic variations have been successfully assessed through the study of correlation and response functions among tree rings and monthly/weekly climatic factors. Moreover, dendrochemical analyses have shown the possibility of detecting the time accumulation of contaminants and pollutants in wood samples [13,34]. Measuring the chemical components concentrations in tree rings has powerful implications both for biological and ecological studies. Information on the spatial/temporal distribution within woody tissues of chemical elements such Ti, Ca and K, which represent basic cellular components or nutrients involved in basic physiological cellular processes, can provide key insights on how trees allocate nutrients in their tissues, e.g., according to species, the age, or other different environmental contexts. In addition, the capability of trees to store in their annual rings both inorganic (e.g., lead) and organic (e.g., HCH) pollutants can provide an a posteriori monitoring tool to quantitatively reconstruct the time dynamics of pollution over a territory [13,17].

In order to identify a preparation procedure for 2D µXRF mapping analysis for the entire core or for a slice about 500 µm thick, we prepared and tested a set of samples from different species, both conifers and angiosperms (figure in Supplementary Materials).

The results point out that together with the expected Ca, Ti and K, we have recorded the presence of high Z elements such as Cr, Fe and Zn. In Figure 4, we show a partial 2D µXRF mapping. The experimental parameters were 100×100 µm^2 steps with 10 s/step acquisition time. The colors (red and blue) representing Ca and Ti and their occurrence in tree rings perfectly reproduced the sequence of different bands of earlywood and latewood, allowing one to position the different elements present as inclusions on a temporal scale.

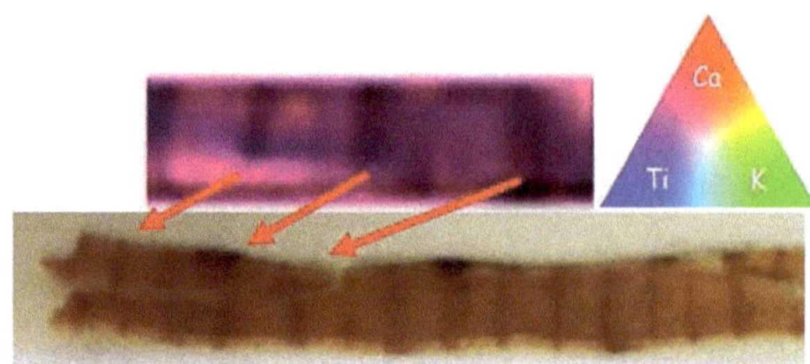

Figure 4. Section of the 2D mapping µXRF. Both Ca (red) and Ti (blue) reproduce the the time sequence of tree rings.

Deep ice cores can provide unaltered records of environmental and climatic changes at different time scales. Moreover, the mineralogical composition as well as the size distribution of particles are related to the dust transport in the atmosphere. Because of this, the dust deposed in the ice cores may provide a record of annual temperature, precipitation, atmospheric composition, volcanic activity and winds [16].

In collaboration with the University of Milan Bicocca and Diamond Lightsources, several samples have been analyzed at XlabF. Antarctic samples previously examined using synchrotron radiation at the Diamond Lightsource and the Stanford Synchrotron Radiation Lightsource (SSRL) have been analyzed, in order to compare the results and to evaluate the mass resolution of our prototype for elemental analysis of low concentration samples [6,7].

As a comparison, in Figure 5, two sets of XRF analysis are reported and compared. For the first set, a sample was deposited on a nuclepore membrane to be analyzed at normal incidence (XRF). For comparison, we reported also the same sample obtained by insoluble dust from Alps ice core (Saharian), analyzed with our prototype and with synchrotron radiation (Diamond Lightsource B-18). Despite the lower intensity on the fluorescence peaks and the higher noise, all elements of interest can also be recognized with our system.

The second example concerns TXRF spectra of a sample of insoluble dust from an Antarctic ice core collected with our prototype and with synchrotron radiation (SSRL Beamline BL-10.2). For both spectra, the measurement conditions were the same: vacuum $\sim 10^{-4}$ hPa for an acquisition time of 600 s. Due to the particular geometry of our system, the detector has to be placed at the Bragg angle ($\sim 40°$) of the Si (400) of the sample and for this reason the silicon diffraction peak (DP) is measured at 7.5 keV. The prototype is able to recognize all the elements of interest in spite of the lower intensity and higher S/N ratio. An upper limit of 10 ng was identified for the interested elements (upper continental crust elements—UCC): Mg, Al, Si, S, Cl, K, Ca, Ti, Mn and Fe. The use of polycapillary semi-lens greatly improves the detection limits of TXRF with a standard source. Although the results remain below that available using an SR source, such a prototype can be used to perform many TXRF researches.

As a summary, Figure 6-top shows the concentrations for both layouts. Using the TXRF layout, the concentration detection is comparable with a synchrotron radiation source, while using the XRF layout our prototype is substantially not comparable with the detection capability of SR, even if samples from the Alps have about one order of magnitude higher concentration with respect to the Antarctic ones. Accordingly, for XRF measurements, our prototype is not suitable for extremely low concentrated samples, such as those from Antarctica. Figure 6-bottom shows the signal-to-noise ratio (SNR) in order to evaluate the limit of detection (LOD), analyzing only the elements in the UCC list. It is important to remark that these LODs are not absolute values, but they are calculated peak to peak for these samples and in a different experimental setup. For TXRF, silicon's SNR is affected by an intrinsic error.

The beam coming from a polyCO semi-lens has millimetric dimensions, so a fraction of the beam contacts the Si-wafer directly.

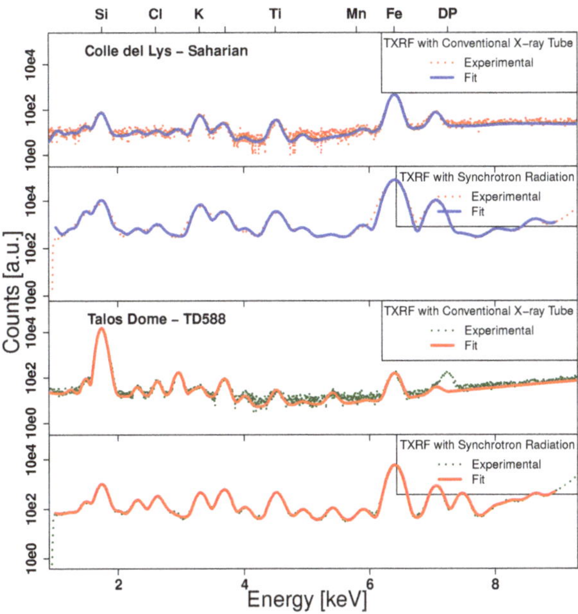

Figure 5. TXRF comparison between our desktop and the Synchrotron Radiation (SR) setup for XRF (first set—Alps sample) and for TXRF (second set—Antarctica sample) measurements.

As a summary, in the top panel of Figure 6, we show the concentrations for both layouts. Using the TXRF layout, the concentration detection is comparable to a synchrotron radiation source, while using the XRF layout the prototype does not compare with the high detection capability of SR. This holds even for Alpine samples with about a one order of magnitude higher concentration with respect to the Antarctic ones. Accordingly, for XRF measurements, this prototype is not suitable for extremely low concentrated samples, such as those from Antarctica. In the bottom panel of Figure 6, to evaluate the limit of detection (LOD) we show the signal-to-noise ratio (SNR) of the elements of the UCC list. It is important to remark here that LODs are not absolute values, but they are values calculated peak to peak for these samples and with this experimental setup. For TXRF, silicon's SNR is affected by an intrinsic error: the beam coming from a polyCO semi-lens has millimeter size dimensions and a fraction of the beam contacts the Si wafer directly. In our sample, magnesium is under the LOD limit, because its energy is near the lower electronic limit of the detector ($\sim$1 keV) and because of the very low concentration—about 2–3% of the sample weight.

For samples from deep ice cores, the preparation technique for XRF measurements is the filtration. The latter presents during the deposition on the membrane some drawbacks such as the coffee stain. Moreover, one of the critical issues is the low superficial density of the deposited dust, highlighted in Figure 6 by the low efficiency of XRF analysis working at normal incidence than synchrotron radiation data. To increase the sample density and the homogeneity of the deposition, new deposition methods based on the micro-drop technique are under test [35]. At present, all samples prepared for TXRF analysis were deposited on Si wafers. However, since the silicon is the element present with the highest concentration in environmental and geological samples, the quantitative analysis of Si with this technique is overestimated. To overcome this issue, other substrates such as lithium fluoride (LiF) crystals are also under test. These preparation methods may certainly enhance the capability of prototypes based on conventional sources with optimized optics if the contaminants present in these substrates are as low as in silicon wafer.

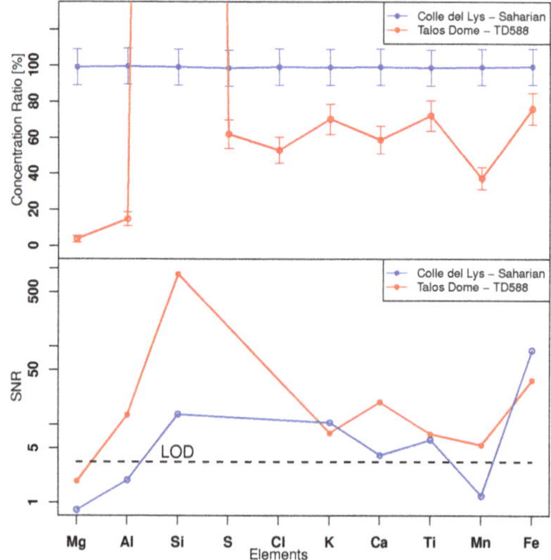

Figure 6. (**Top**) Concentration ratio for the samples in Figure 5. In the TXRF layout, a comparable concentration detection is achieved, while for the XRF modality the elements concentrations are not comparable. (**Bottom**) The signal-to-noise ratio (SNR) is analyzed in order to evaluate the limit of detection (LOD) of our system with these samples. The elements analyzed are the only in the upper continental crust elements (UCC) list. The Si element has an intrinsic error: the beam coming from a polyCO semi-lens has millimetric dimensions. Thus, a fraction of the beam directly contacts the Si wafer.

4. Conclusions

The peculiar capability of polycapillary optics to obtain a powerful quasi-parallel beam from a divergent source provides great opportunities for elemental analysis by using both XRF and TXRF even for very low concentration samples. The comparison with synchrotron radiation data shows that the results presented in this contribution are particularly interesting. Indeed, data of two independent sets show good agreement with experimental data collected using SR. The agreement is particularly good at energies up to the Fe K-emission line, indicating that a conventional X-ray source combined with polycapillary optics is a valuable alternative for many analysis in samples with elemental concentrations down to ng/g. Best results were obtained with the TXRF layout. However, this technique requires a sample preparation to be optimized for geological researches, particularly regarding the substrate and the sample deposition method. Currently, we are evaluating two different types of substrates such as LiF crystals and polymeric films. However, the final choice must be to pay particular attention to the presence of contaminants in the substrates. Work is also in progress to optimize the deposition method and in particular to increase the sample density and the deposition homogeneity.

Supplementary Materials: Supplementary Materials can be found at: http://www.mdpi.com/2410-3896/3/4/33/s1.

Author Contributions: D.H., S.B.D., G.C. and A.M. are responsible of the project and realization of the RXR experimental setup and development of the control software, V.M is the team leader of the Antarctic and Alps drilling sites, A.d.F. and G.P. provided the tree samples. All authors have read and approved the final manuscript.

Funding: This research received no external funding.

Acknowledgments: One of the authors (S.B.D.) acknowledges the support by the Competitiveness Program of NRNU MEPhI. In collaboration with the Laboratory of Dendroecology of the Department of Agriculture and Forest Science (DAFNE) of the University of Tuscia, we started at XlabF to develop a protocol to analyze tree-ring cores necessary to monitor beech forests patronaged by UNESCO. In collaboration with the University of Milan Bicocca and Diamond Lightsources, several samples have been analyzed using synchrotron radiation at the Diamond Lightsource and the Stanford Synchrotron Radiation Lightsource (SSRL).

Conflicts of Interest: The authors declare no conflict of interest.

Abbreviations

The following abbreviations are used in this manuscript:

PolyCO	Polycapillary Optics
XRF	X-ray Fluorescence
µXRF	micro X-ray Fluorescence
TXRF	Total Reflection X-ray Fluorescence
SR	Synchrotron Radiation
UCC	Upper Continental Crust

References

1. Tsuji, k.; Nakano, K.; Hayashi, H.; Hayashi, K.; Ro, C. X-ray Spectrometry. *Anal. Chem.* **2008**, *80*, 4421–4454. [CrossRef] [PubMed]

2. Hampai, D.; Dabagov, S.B.; Cappuccio, G.; Longoni, A.; Frizzi, T.; Cibin, G.; Guglielmotti, V.; Sala, M. Elemental Mapping and Micro-Imaging by X-Ray Capillary Optics. *Opt. Lett.* **2008**, *33*, 2743–2745. [CrossRef] [PubMed]

3. Giannoncelli, A. In Proceedings of the 24th International Congress on X-Ray Optics and Microanalysis, Trieste, Italy, 25–29 September 2017.

4. Croudace, I.W.; Rothwell, R.G. *MIcro-XRF Studies of Sediment Cores*; Springer: Berlin, Germany, 2015.

5. Shackley, M.S. *X-ray Fluorescence Spectrometry (XRF) in Geoarchaeology*; Spinger: Berlin, Germany, 2011.

6. Cibin, G.; Marcelli, A.; Maggi, V.; Sala, M.; Marino, F.; Delmonte, B.; Albani, S.; Pignotti, S. First combined total reflection X-ray fluorescence and grazing incidence X-ray absorption spectroscopy characterization of aeolian dust archived in Antarctica and Alpine deep ice cores. *Spectrochim. Acta B* **2008**, *63*, 1503–1510. [CrossRef]

7. Marcelli, A.; Cibin, G.; Hampai, D.; Giannone, F.; Sala, M.; Pignotti, S.; Maggi, V.; Marino, F. XANES characterization of deep ice core insoluble dust in the ppb range. *J. Anal. At. Spectrom.* **2012**, *22*, 33–37. [CrossRef]

8. Dabagov, S.B. Channeling of neutral particles in micro- and nanocapillaries. *Phys. Uspekhi* **2003**, *46*, 1053–1075. [CrossRef]

9. Hampai, D.; Cherepennikov, Y.M.; Liedl, A.; Cappuccio, G.; Capitolo, E.; Iannarelli, M.; Azzutti, C.; Gladkikh, Y.P.; Marcelli, A.; Dabagov, S.B. Polycapillary based µXRF station for 3D colour tomography. *JINST* **2018**, *13*, C04024. [CrossRef]

10. Hampai, D.; Liedl, A.; Cappuccio, G.; Capitolo, E.; Iannarelli, M.; Massussi, M.; Tucci, S.; Sardella, R.; Sciancalepore, A.; Polese, C.; et al. 2D-3D µXRF elemental mapping of archeological samples. *Nucl. Instrum. Methods B* **2017**, *402*, 274–277. [CrossRef]

11. Zoeger, N.; Streli, C.; Wobrauschek, P.; Jokubonis, C.; Pepponi, G.; Roschger, P.; Hofstaetter, J.; Berzlanovich, A.; Wegrzynek, D.; Chinea-Cano, E.; et al. Determination of the elemental distribution in human joint bones by SR micro XRF. *X-Ray Spectrom.* **2008**, *37*, 3–11. [CrossRef]

12. Nakano, K.; Tanaka, K.; Ding, X.; Tsuji, K. Development of a new total reflection X-ray fluorescence instrument using polycapillary X-ray lens. *Spectrochim. Acta B* **2006**, *61*, 1105–1109. [CrossRef]

13. Bernini, R.; Pelosi, C.; Carastro, I.; Venanzi, R.; Di Filippo, A.; Piovesan, G.; Ronchi, B.; Danieli, P.P. Dendrochemical investigation on hexachlorocyclohexane isomers (HCHs) in poplars by an integrated study of micro-Fourier transform infrared spectroscopy and gas chromatography. *Trees Struct. Funct.* **2016**, *30*, 1455–1463. [CrossRef]

14. Petit, J.R.; Jouzel, J.; Raynaud, R.; Barkov, N.I.; Barnola, J.M.; Basile, I.; Bender, M.; Chappellaz, J.; Davis, M.; Delaygue, G.; et al. Climate and atmospheric history of the past 420,000 years from the Vostok ice core, Antarctica. *Nature* **1999**, *399*, 429–436. [CrossRef]

15. Community Members EPICA. Eight glacial cycles from an Antarctic ice core. *Nature* **2004**, *429*, 623–628. [CrossRef] [PubMed]

16. Maggi, V. Mineralogy of atmospheric microparticles deposited along the Greenland Ice Core Project ice core. *J. Geophys. Res.* **1997**, *102*, 26725–26734. [CrossRef]

17. Sawidis, T.; Breuste, T.; Mitrovic, M.; Pavlovic, P.; Tsigaridas, K. Trees as bioindicator of heavy metal pollution in three European cities. *Environ. Pollut.* **2011**, *159*, 3560–3570. [CrossRef] [PubMed]

18. Hampai, D.; Dabagov, S.B.; Cappuccio, G. Advanced studies on the Polycapillary Optics use at XLab Frascati. *Nucl. Instrum. Methods B* **2015**, *355*, 264–267. [CrossRef]

19. Nanoray, a portable X-ray machine. FP7 Project N 222426 (2008-2011). Available online: https://www.parlementairemonitor.nl/9353000/1/j9tvgajcor7dxyk_j9vvij5epmj1ey0/vj4871mm6efo?ctx=vg9pjpw5wsz1&start_tab0=1160 (accessed on 18 October 2018).

20. Hampai, D.; Dabagov, S.B.; Cappuccio, G.; Cibin, G.; Sessa, V. X-ray micro-imaging by capillary optics. *Spectrochim. Acta Part B* **2009**, *64*, 1180–1184. [CrossRef]

21. Cherepennikov, Y.; Miloichikova, I.; Gogolev, A.; Stuchebrov, S.; Hampai, D.; Dabagov, S.; Liedl, A. Application of polycapillary optics for dual energy spectroscopy based on a laboratory source. *Nucl. Instrum. Methods B* **2017**, *402*, 278–281. [CrossRef]

22. Liedl, A.; Dabagov, S.B.; Hampai, D.; Polese, C.; Tsuji, K. On X-ray channeling in a vibrating capillary. *Nucl. Instrum. Methods B* **2015**, *355*, 289–292. [CrossRef]

23. Marchitto, L.; Hampai, D.; Dabagov, S.B.; Allocca, L.; Alfuso, S.; Polese, C.; Liedl, A. GDI spray structure analysis by polycapillary X-ray μ-tomography. *Int. J. Multiph. Flow* **2015**, *70*, 15–21. [CrossRef]

24. Bonfigli, F.; Hampai, D.; Dabagov, S.B.; Montereali, R.M. Characterization of X-ray polycapillary optics by LiF crystal radiation detectors through confocal fluorescence microscopy. *Opt. Mater.* **2016**, *58*, 398–405. [CrossRef]

25. Gogolev, A.S.; Hampai, D.; Khusainov, A.K.; Zhukov, M.P.; Dabagov, S.B.; Potylitsyn, A.P.; Liedl, A.; Polese, C. Results of testing the energy dispersive Si detector with large working area. *Nucl. Instrum. Methods B* **2015**, *355*, 268–271. [CrossRef]

26. Available online: http://www.oxford-instruments.com (accessed on 18 October 2018).

27. Available online: http://www.xglab.it (accessed on 18 October 2018).

28. Available online: http://www.photonic-science.com (accessed on 18 October 2018).

29. Available online: https://www.physikinstrumente.com/en/ (accessed on 18 October 2018).

30. Available online: http://www.axo-dresden.de/products/highprecision/reference.htm (accessed on 18 October 2018).

31. Bostick, B.C.; Theissen, K.M.; Dunbar, R.B.; Vairavamurthy, A. Record of redox status in laminated sediments from Lake Titicaca: A sulfur K-edge X-ray absorption near edge structure (XANES) study. *Chem. Geol.* **2005**, *219*, 163–174. [CrossRef]

32. Moy, C.M.; Dunbar, R.B.; Guilderson, T.P.; Waldmann, N.; Mucciarone, D.A.; Recasens, C.; Ariztegui, D.; Austin, J.A., Jr.; Anselmetti, F.S. A geochemical and sedimentary record of high southern latitude Holocene climate evolution from Lago Fagnano, Tierra del Fuego. *Earth Planet. Sci. Lett.* **2011**, *302*, 1–13. [CrossRef]

33. Polyak, V.J.; Asmerom, Y. Late Holocene Climate and Cultural Changes in the Southwestern United States. *Science* **2001**, *294*, 148–151. [CrossRef] [PubMed]

34. Di Filippo, A.; Biondi, F.; Cûfar, K.; De Luis, M.; Grabner, M.; Maugeri, M.; Presutti Saba, E.; Schirone, B.; Piovesan, G. Bioclimatology of beech (*Fagus sylvatica* L.) in the Eastern Alps: Spatial and altitudinal climatic signals identified through a tree-ring network. *J. Biogeogr.* **2007**, *34*, 1873–1892. [CrossRef]

35. Macis, S.; Cibin, G.; Maggi, V.; Baccolo, G.; Hampai, D.; Delmonte, B.; D'Elia, A.; Marcelli, A. Microdrop Deposition Technique: Preparation and Characterization of Diluted Suspended Particulate Samples. *Condens. Matter* **2018**, *3*, 21. [CrossRef]

Article

Microdrop Deposition Technique: Preparation and Characterization of Diluted Suspended Particulate Samples

Salvatore Macis [1,2,*], Giannantonio Cibin [3], Valter Maggi [4,5], Giovanni Baccolo [4,5],
Dariush Hampai [2], Barbara Delmonte [4], Alessandro D'Elia [6] and Augusto Marcelli [2,7]

1. Department of Mathematics and Physics, Università di Roma Tor Vergata, via della Ricerca Scientifica 1, 00133 Rome, Italy
2. Laboratori Nazionali di Frascati, Istituto Nazionale di Fisica Nucleare, 00044 Frascati, Italy; dariush.hampai@lnf.infn.it (D.H.); augusto.marcelli@lnf.infn.it (A.M.)
3. Diamond Light Source, Harwell Science and Innovation Campus, Didcot OX11 0DE, UK; giannantonio.cibin@diamond.ac.uk
4. Dipartimento di Scienze dell'Ambiente e della Terra, Università degli Studi di Milano Bicocca, Piazza della Scienza, 1-20126 Milano, Italy; valtermaggi@gmail.com (V.M.); giovanni.baccolo@mib.infn.it (G.B.); barbara.delmonte@unimib.it (B.D.)
5. Sezione di Milano-Bicocca, Istituto Nazionale di Fisica Nucleare, Piazza della Scienza, 3-20126 Milano, Italy
6. Department of Physics, University of Trieste, Via A. Valerio 2, 34127 Trieste, Italy; ale9149@gmail.com
7. Istituto Officina dei Materiali, Consiglio Nazionale delle Ricerche, Basovizza SS-14, km 163.5, 34149 Trieste, Italy
* Correspondence: salvatore.macis@roma2.infn.it; Tel.: +39-067-259-4523

Received: 23 May 2018; Accepted: 11 July 2018; Published: 16 July 2018

Abstract: The analysis of particulate matter (PM) in dilute solutions is an important target for environmental, geochemical, and biochemical research. Here, we show how microdrop technology may allow the control, through the evaporation of small droplets, of the deposition of insoluble materials dispersed in a solution on a well-defined area with a specific spatial pattern. Using this technology, the superficial density of the deposited solute can be accurately controlled. In particular, it becomes possible to deposit an extremely reduced amount of insoluble material, in the order of few µg on a confined area, thus allowing a relatively high superficial density to be reached within a limited time. In this work, we quantitatively compare the microdrop technique for the preparation of particulate matter samples with the classical filtering technique. After having been optimized, the microdrop technique allows obtaining a more homogeneous deposition and may limit the sample amount up to a factor 25. This method is potentially suitable for many novel applications in different scientific fields such as demanding spectroscopic studies looking at the mineral fraction contained in ice cores or to pollution investigations looking at the detection of heavy metals present in ultra-trace in water.

Keywords: ultra-dilution; droplets; water; evaporation; X-ray fluorescence

---

1. Introduction

When dealing with complex samples, one of the main bottlenecks from the practical point of view is their preparation. A suitable sample preparation method should allow saving time during data acquisition, enhancing the S/N ratio, improving the detection limits, etc. Considering particulate matter (PM) analysis of solutions, the most common preparation technique is filtration: it is easy, reliable, and fast, but at the same time, it presents some drawbacks. Indeed, it requires a considerable sample consumption and PM whose size is smaller than the one of the filtration pores, which are not retained on the filter. To avoid this point, the original solution could be deposited on a membrane presenting a quite large surface (about 1 cm^2). In this case a critical issue is the low superficial density of the PM deposited on top of it.

The evaporation process, based on the deposition of micro-droplets, may overcome both issues. Indeed, the microdrop technique consists in the deposition of PM in association to liquid drops presenting a diameter between 70 and 200 μm. After the evaporation of the solvent, the PM depositional pattern resembles the one of the original droplets. The main disadvantages of this method are the inhomogeneity of the deposition and the long time required for the evaporation of the solvent. In fact, the evaporation of large drops (few mm of diameter) can last several hours. At normal conditions of temperature and humidity, full evaporation of drops of water can require more than 10 h per mL, and the density fluctuation of the deposited material is up to 80% as a consequence of the so-called coffee-stain effect [1]. To optimize the process, we deposit micro-droplets and control the deposition area with motorized translational stages. To this purpose, a study of the evaporation dynamics is necessary to investigate the uniformity of deposition and to evaluate the evaporation rate of micro-droplets presenting different sizes. When dealing with densely-packed deposition patterns, our results show that is necessary to consider the interaction between different evaporating droplets.

The evaporation of a drop on a surface can be described with two different models: the evaporation with a "constant contact area" and that with a "constant contact angle" [2] (see Figure 1).

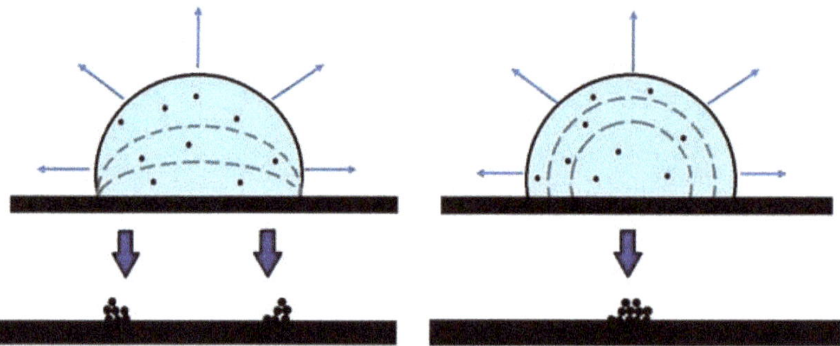

Figure 1. Deposition and evaporation processes for two different models: "constant contact area" (left) and "constant contact angle" (right). The main difference is represented by the shape of materials deposited on the substrate: pinned at the edge for the "constant contact area" model (bottom left) and concentrated in the center for the "constant contact angle" model (bottom right).

In the first case, the evaporation takes place, maintaining a constant contact area between the liquid drop and the surface. The shape of the drop remains almost "spherical" [2], but the contact angle decreases. In accordance to the second model, the contact angle of the edge of the drop is constant during the evaporation [2], thus, the shape of the drop remains "spherical", but the contact area between liquid and surface continuously decreases. The "constant contact area model" is suitable when a strong interaction between the liquid and the substrate takes place, e.g., when considering water and a hydrophilic surface [3,4]. This is clear in the work of Peschel et al., where at higher concentration, the evaporation occurs with the constant area mode, i.e., during the evaporation, the droplet does not shrink [5].

On the opposite, the second model is more appropriate when the drop-substrate interactions are weak, as in the case of water on a hydrophobic substrate. As studied by Fittschen et al. [6], an ultra-diluted solution is characterized by a low solution–substrate interaction, and the coffee stain phenomenon is reduced. As a matter of fact, the insoluble components deposed by the evaporation of the droplets have to be characterized in order to achieve the most homogeneous deposition starting from defined droplet size. As shown by Sparks et al., droplet dimension and solution concentration are the main parameters that control the deposition [7]. Thus, the deposition method we presented can be a fast, simple, and direct way to achieve a homogeneous deposition of insoluble materials considering only the droplet size and the concentration of the solution.

In this contribution, we will describe the optimization process of an experimental setup based on the evaporation of water micro-droplets in accordance to the constant angle model. The uniformity of the deposition will be determined, and results will be compared to those obtained through the standard filtration method. Indeed, the microdrop method may offer many opportunities and dedicated preparations in different scientific fields. One particular demanding application is the homogeneous deposition onto specific substrates of inorganic components, e.g., coarse particles, mineral fractions, and particulate matter for spectroscopic characterizations. Research that could benefit from this particular type of sample preparation are ice core studies or detection of heavy metals present in ultra-trace elements.

2. Experimental

Using a micro-dispenser, micro-sized droplets can be deposited, controlling the spatial distribution of drops, their size, and the deposition rate. The Microdrop technology [8] allows spreading of extremely small amounts of a liquid solution. Moreover, thanks to dedicated devices, it is possible to deposit single droplets whose volume is in the sub-nanoliter range. Such small droplets present an interaction with the substrate, which can be considered negligible. The micro-dispenser is composed of a head with a nozzle and a pumping chamber working with the same piezoelectric technology used for inkjet printers. The head is controlled by a driver, i.e., a pulse generator, which controls the piezoelectric actuator and sends pulses to the head. The device we used is the MD-K-130 (© Microdrop Technologies GmbH, Norderstedt, Germany). With an inner nozzle diameter of 70 μm, we may produce, with 1% volume repeatability, droplets with a minimum nominal volume of 180 pl and at a drop rate from 1 to 2000 Hz [8,9]. To prepare a sample with a specific microdrop pattern, it is necessary to control the number of drops on the substrate, releasing each of them at a well-defined and repeatable distance. In our experimental setup, to generate the pattern, the head is maintained fixed while the substrate translates under the stream of droplets released by the head. The motion of the substrate is realized with two precision PI Micos (Physik Instrumente GmbH Karlsruhe, Karlsruhe, Germany) Translational Stage VT-80 stages, assembled perpendicularly with respect to each other. This setting allows positioning the substrate with a nominal accuracy of 1 μm per axis. The desired spatial patterns are obtained controlling the two translation stages with a LabView©-based code and properly setting the Microdrop dispenser [9].

The characterization of the evaporation time and of the drop pattern was made with the Zeiss Axio Imager M1 optical microscope (Carl Zeiss Microscopy LLC, Oberkochen, Germany) using 5×, 10×, and 20× magnification optics. The video camera had a x–y spatial resolution of 0.13 μm at 5×, 0.26 μm at 10×, and 0.52 μm at 20× magnification and allowed us to measure the droplet volume with an uncertainty of ~3% of the measured volume. We used a solution of ink and water (Gullor® ink diluted at 50% with high purity milliQ water) for the optical analysis of the deposition, and high purity milliQ water for the evaporation dynamic measurements, i.e., the diameter of the drop and the angle of contact. The substrates we used for the evaporation analysis were Kapton® polyimide films from DuPont of 0.005′ (125 μm) thickness, while we used polycarbonate Nucleopore© membrane filters (pore size 0.45 μm and filter area diameter ~20 mm) for the analysis of morphology on evaporated samples morphology by X-Ray Fluorescence.

The evaluation of the PM deposition uniformity has been performed using the X-ray fluorescence technique at the beamline B18 at the *Diamond Light Source* facility (Harwell, United Kingdom) [10], using a monochromatic X-ray beam at 8 keV, with a focus of 100 μm radius and a detection limit of 100 ng/mm^2.

The evaluation of the evaporation was carried out measuring the diameter and height of droplets as a function of time using the optical microscope and the camera. We measured the parameters of the single droplets and of groups of droplets to evaluate the evaporation rate as a function of the size and of the distance between several droplets. We then identified the minimum deposited area achievable with the evaporation of small droplets, in the shortest time. To optimize the deposition process, all images and experimental parameters were analyzed and used to write an approximate model.

3. Results

The analysis of the images of the droplets (see Figure 2) and of the measured parameters clearly points out that when using a low concentration solution and a hydrophobic substrate such as *Kapton*, the evaporation occurs in accordance to the "constant angle" model, well described by the theoretical model introduced by Picknett and Bexon in 1977 [2]. The result is confirmed in Figure 3, where our data are compared to the "constant angle" and "constant contact area" models. The deposition rate, i.e., the volume of the liquid deposited vs. time, was optimized considering the velocity of the stages and the spatial pattern parameters of the deposited droplets. Indeed, a fundamental parameter that drives the evaporation is the saturation of the atmosphere surrounding the droplets' selves, which is directly related to the mutual position of the droplets. Different droplet sizes and different distances between successive droplets were considered to this aim. In Figure 4, we show the correlated parameters of the deposition tests we performed using the translation stages combined with the microdrop. We defined the area to cover (simulating an experimental condition, e.g., the deposition of 1 mL of solution onto an area of ~35 mm^2) and then showed data using the distance between two consecutive drops along the deposition line, and the diameter of the droplet to be deposited as the two axes. The third axis represents the evaporation rate density, i.e., the amount that can be deposited on the surface area (defined above) per unit/area. Data presented in Figure 4 point out that the highest evaporation rate at 20 °C (~0.7 mL/h cm^2) is obtained setting the droplet diameter at 240 μm and the distance between two consecutive droplets at 170 μm.

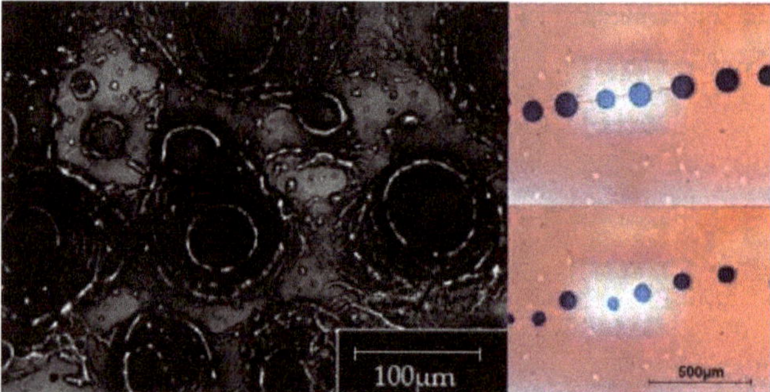

Figure 2. Optical microscope image taken at 20× magnification of the microdrop deposition of multiple 180 pl droplets with the ink solution deposed on the kapton substrate (left). On the right two microscope images taken at 5× magnification of a line of distilled water droplets: one at the beginning of the evaporation (top right) and the second after 10 s (bottom right).

Closer drops would determine a higher relative humidity in the air layers, which surround the drops selves, slowing the evaporation process. At variance, depositing drops at larger distance will limit the amount of deposited liquid actually reducing the integrated efficiency of this process. To further increase the evaporation rate, experiments conducted heating the substrate were also carried out. As expected, as temperature increases, the evaporation rate increases exponentially, with exponential value of the temperature of ~2.5. For example, at 80 °C, the evaporation rate increases about 30 times [9]. However, it is worth noting that increasing temperature could not be always feasible, since increasing the temperatures may affect the chemical stability of the samples.

To avoid the problem, thermal radiation focused on droplets may be considered. By selecting specific wavelengths or frequencies, it would be possible to selectively increase the temperature of the solvent and not the one of the PM contained in the solution. Another parameter, which could be modified to fasten the deposition, is the speed of the two motorized stages. Our experiments were all conducted at the maximum allowed speed (i.e., 13 mm·s^{-1}), but faster motorized stages exist that would cover the same area within a shorter time, i.e., just reducing the deposition time between two consecutive drops.

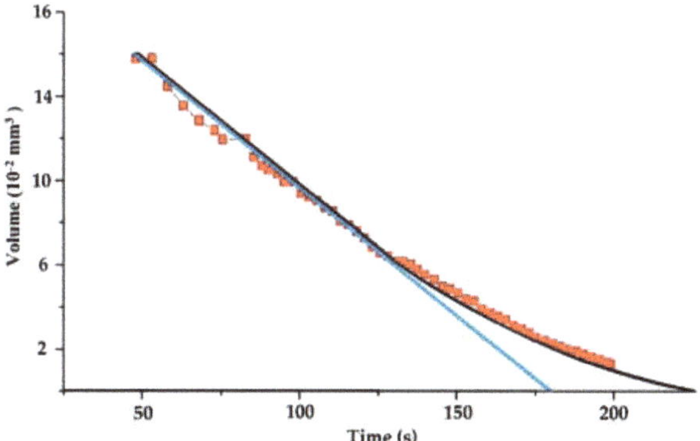

Figure 3. Comparison among the droplet mass behavior for the evaporation at "constant contact area" (blue line), at "constant contact angle" (black line), and experimental data (red squares) of a droplet of a bi-distilled water solution with an initial diameter of 1 mm, at room temperature.

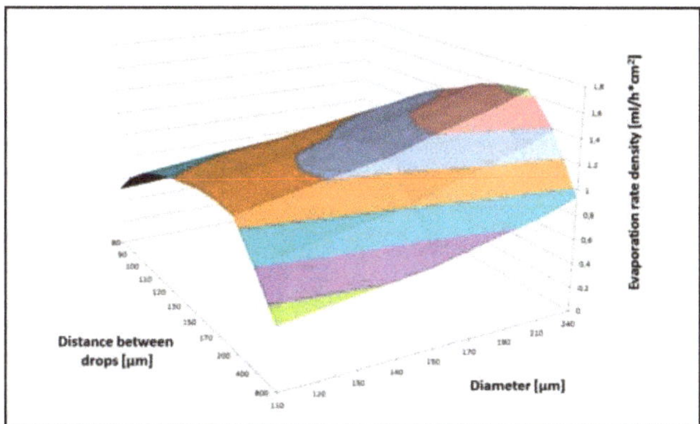

Figure 4. This graph shows the evaporation rate density (volume evaporated per hour and area) at 20 °C vs. drop diameter and distance between two consecutive drops. Here, the speed has also been considered, i.e., the time associated with the motion of the translational stages used to map with droplets the area to cover.

To better understand the deposition pattern of the PM contained in a liquid solution deposited with a Microdrop dispenser, we prepared a suspension consisting in high purity water (MilliQ technology) and a given amount (50 μg/mL) of a reference material: the NIST standard 2709a. The latter is a well-known soil reference material with a well-characterized composition [11] and grain size distribution. To avoid the clog of the Microdrop nozzle, we selected a reference material whose particles didn't exceed 30 μm, i.e., about the half of the aperture of the nozzle orifice, as shown in Figure 5.

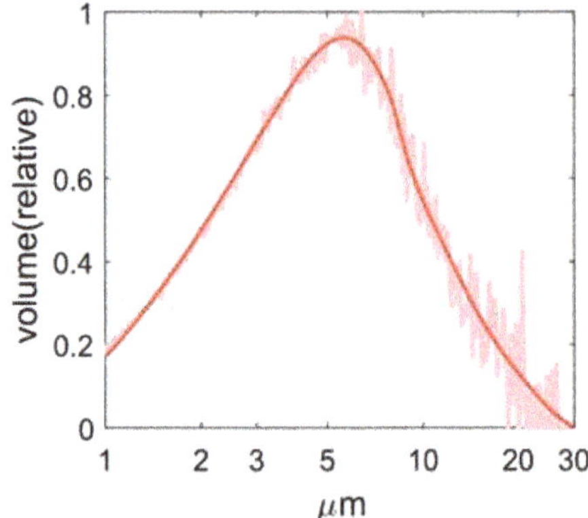

Figure 5. The particle size distribution of the NIST standard reference soil material 2709a. Data were obtained through the Coulter counter technique. Details can be found in [12].

Since the standard contains Fe, it was possible to use the X-ray fluorescence (XRF) technique to investigate its spatial distribution after deposition. In Figure 6, we compare two XRF profiles collected on two deposited samples, measured along a straight line, with a spatial resolution limited by the spot size of the beam (~200 μm). The red curve is the profile relative to the deposition of a single large drop of ~1 mL volume of the NIST solution. The black curve refers to a deposition obtained with the microdrop, made by multiple patterns of 240 μm diameter droplets with 170 μm between them. The same total volume and the same solution were considered. The two profiles show significant differences and it is remarkable that the homogeneous deposition was obtained with the latter technique, despite the reduced amount of solution. Indeed, the large drop shows density fluctuations from 50% to 80% of the fluorescence intensity, in particular at the edge due to the coffee-stain effect [1]. At variance, the variability observed along the microdrop profile is around 10%, and the deposited soil is uniformly distributed.

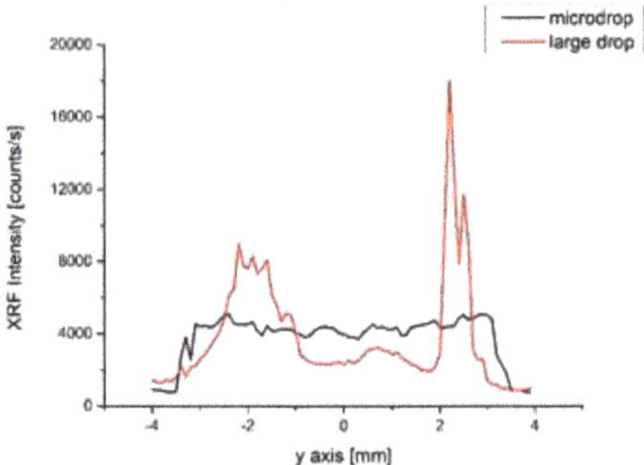

Figure 6. Comparison of two representative X-ray fluorescence profiles at the Fe K-edge: microdrop 5×7 mm^2 pattern with drops of 2.9 μL (240 μm diameter—black line) and a single large drop deposition (red line). Both samples have been obtained with a total volume deposition of 1 ± 0.1 mL of a NIST solution 50 μg/mL concentration on a kapton film.

Figure 6 compares a microdrop deposition, which clearly shows the homogeneity and the overall quality of the samples prepared with the microdrop increase. The only disadvantage of this method is the deposition time. Indeed, as it can be inferred from Figure 4, the optimal configuration of this micro-deposition allows us to deposit and to evaporate liquid samples with a rate of less than 1 mL per hour. A simple and alternative way to reduce the deposition time can be obtained by combining the microdrop deposition with a simultaneous filtering technique. In this way, the droplet is deposited onto a wettable filter, and the liquid is pumped out by a vacuum pump.

In Figure 7 (left), the 2 × 5 mm² image shows a portion of the original 4 × 4 mm² microdrop deposition area of 1 mL NIST solution (50 µg/mL concentration) on a polycarbonate Nucleopore© membrane filter (pore size 0.45 µm and filter area diameter ~20 mm), which is a hydrophilic substrate. This type of substrate absorbs the liquid solution, and the liquid is continuously pumped out through the filter. The color map was obtained collecting the XRF signal at the Fe K-edge. Colors highlight an area of approximately 2 × 3 mm² where a controlled distribution is achieved. The internal variability is probably also related to the spatial pattern of the pores characterizing the filtration membrane.

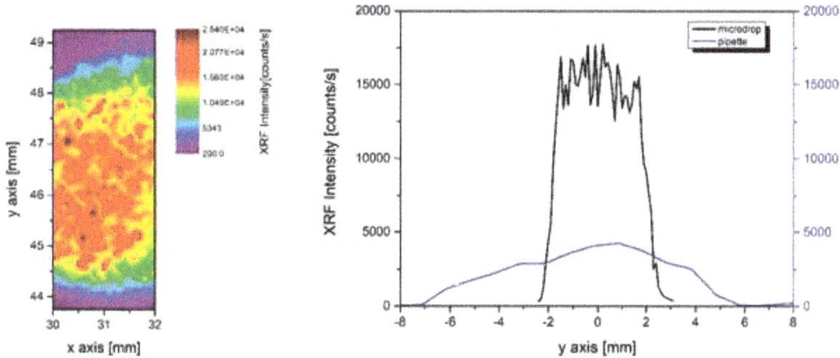

Figure 7. Image of the X-ray Fluorescence at the Fe K-edge of the microdrop deposition area on the filter (left); comparison of two XRF profiles (right): the microdrop deposition (black) and the pipette deposition (blue) both achieved with a simultaneous vacuum filtration technique.

In Figure 7 (right), one microdrop deposition profile of the image on the left is compared with an analogous profile of the deposition of the same solution using a micro-pipette on the same type of filter. From the analysis of the data, we may point out that by using the microdrop deposition, ~80% of the material is distributed quite homogeneously (with a fluorescence intensity fluctuation of ~15%) inside an area of ~8 mm². Using the pipette deposition, the same amount of material is deposited within an area six times larger (~50 mm²). A simple evaluation points out that the microdrop filtration procedure covers an area 6.25 times smaller than the area wetted with the pipette, but the intensity of the fluorescence signal, proportional to the deposed material is four times higher. In this case, the total gain we achieved is ~ 25.

4. Conclusions

The microdrop technology is suitable to prepare homogeneous deposition of particulate and granular matter samples. In order to establish a reliable procedure, reducing also the deposition time, it is necessary to measure accurately the evaporation rate of the pattern depositions. We show here how to optimize such parameters, increasing the S/N ratio and the deposition homogeneity, reducing at the same time the sample consumption. Indeed, S/N ratio is related to the XRF signal; increasing the density and the homogeneity of the deposition, the S/N ratio of the XRF signal that outlines the distribution of the material in the sample improves. In terms of deposition uniformity, with the microdrop technique, the S/N ratio may improve ~4 times with respect to a classical filtration method. At the same time, it is possible to deposit and evaporate a solution onto any substrate and 5 times more homogeneously than a large drop evaporation. These features make this technique effective and competitive with respect to many and diverse applications. Among the many, we cite the characterization of aerosols for pollution monitoring purposes, studies of metallic contaminants in polluted water or the investigation of biological diluted materials at ultra-trace concentrations.

Besides homogeneity, another important feature of the method proposed here, is the possibility to concentrate in a very reduced area the material contained in an extremely diluted solution or suspension. This makes it possible to consider impurities whose concentrations span from few ppb to hundreds ppm. As an example, the method is suited for the characterization of the inorganic insoluble particulate matter contained in ice core samples for paleoclimatic reconstructions. Ice core dust samples prepared with a microdrop device could be successfully used for both X-rays absorption and X-ray fluorescence measurements [13,14] and under particular experimental conditions also for XRD experiments [15]. Certainly, the microdrop technique has to be improved for applications where beams with spot size well below one micron will be available, as to ultra low brilliance synchrotron radiation facilities and/or Free Electron Lasers. However, due to the low frequency rate of the existing Free Electron Lasers (10–100 Hz), the technique could already be useful to control the deposition of small samples such as small crystals, clusters, or organic systems delivered to the beam inside droplets.

Author Contributions: S.M., G.C., V.M. and A.M. conceived and designed the experiment; D.H. and S.M. built the experimental setup; S.M. and G.C. performed the experiments and analyzed data; G.B., B.D. and A.D. provided materials and contributed to the data analysis; S.M., G.C., and A.M. wrote the paper. All authors have read and approved the final manuscript.

Funding: This research has been performed in cooperation among INFN-Laboratori Nazionali di Frascati, Milano Bicocca University and the Diamond Light Source facility. The development of the microdrop instrumentation is a project managed by A.M. with the support of DARST—Department of the Presidenza del Consiglio dei Ministri, which is gratefully acknowledged. Part of the preparation was done in the EuroCold Laboratory at University of Milano Bicocca, founded by Italian National Science Foundation NextData Project. Part of this research was developed at Diamond, the UK national synchrotron radiation facility. One of us (S.M.) acknowledges the Roma Tre University for financial support during his stage at Diamond.

Conflicts of Interest: The authors declare no conflict of interest.

References

1. Innocenzi, P.; Malfatti, L.; Piccinini, M.; Grosso, D.; Marcelli, A. Stain Effects Studied by Time-Resolved Infrared Imaging. *Anal. Chem.* **2009**, *81*, 551–556. [CrossRef] [PubMed]

2. Picknett, R.G.; Bexon, R. The evaporation of sessile or pendant drops in still air. *J. Colloid Interface Sci.* **1977**, *61*, 336–350. [CrossRef]

3. Hu, H.; Larson, R.G. Evaporation of a sessile droplet on a substrate. *J. Phys. Chem. B* **2002**, *106*, 1334–1344. [CrossRef]

4. Girard, F.; Antoni, M.; Faure, S.; Steinchen, A. Evaporation and Marangoni Driven Convection in Small Heated Water Droplets. *Langmuir* **2006**, *22*, 11085–11091. [CrossRef] [PubMed]

5. Peschel, B.U.; Fittschen, U.E.A.; Pepponi, G.; Jokubonis, C.; Streli, C.; Wobrauschek, P.; Falkenberg, G.; Broekaert, J.A.C. Direct analysis of Al_2O_3 powders by total reflection X-ray fluorescence spectrometry. *Anal. Bioanal. Chem.* **2005**, *382*, 1958–1964. [CrossRef] [PubMed]

6. Fittschen, U.E.A.; Bings, N.H.; Hauschild, S.; Förster, S.; Kiera, A.F.; Karavani, E.; Frömsdorf, A.; Thiele, J.; Falkenberg, G. Characteristics of picoliter droplet dried residues as standards for direct analysis techniques. *Anal. Chem.* **2008**, *80*, 1967–1977. [CrossRef] [PubMed]

7. Sparks, C.M.; Fittschen, U.E.A.; Havrilla, G.J. Picoliter solution deposition for total reflection X-ray fluorescence analysis of semiconductor samples. *Spectrochim. Acta Part B* **2010**, *65*, 805–811. [CrossRef]

8. Meyer, W. *Mikroverklebungen aus dem Tintenstrahldrucker PLUS*; Eugen, G., Ed.; Leuze Verlag: Norderstedt, Germany, 2013; pp. 593–598. (In German)

9. Macis, S. Preparation and Characterization of Ultra-Diluted Samples via Micro-Deposition of Droplets. Master's Thesis, Roma Tre University, Roma, Italy, 2014.

10. Dent, J.; Cibin, G.; Ramos, S.; Parry, S.A.; Gianolio, D.; Smith, A.D.; Scott, S.M.; Varandas, L.; Patel, S.; Pearson, M.R.; et al. Performance of B18, the Core EXAFS Bending Magnet beamline at Diamond. *J. Phys. Conf. Ser.* **2013**, *430*, 012023. [CrossRef]

11. Mackey, E.A.; Christopher, S.J.; Lindstrom, R.M.; Long, S.E.; Marlow, A.F.; Murphy, K.E.; Paul, R.L.; Popelka-Filcoff, R.S.; Rabb, S.A.; Sieber, J.R.; et al. Nebelsick, *Certification of Three NIST Renewal Soil Standard Reference Materials for Element Content: SRM 2709a San Joaquin Soil, SRM 2710a Montana Soil I, and SRM*

2711a *Montana Soil II*; NIST Special Publication 260-172; National Institute of Standards and Technology: Gaithersburg, MD, USA, 2010; p. 39.

12. Ruth, U.; Barbante, C.; Bigler, M.; Delmonte, B.; Fischer, H.; Gabrielli, P.; Gaspari, V.; Kaufmann, P.; Lambert, F.; Maggi, V.; et al. Proxies and Measurement Techniques for Mineral Dust in Antarctic Ice Cores. *Environ. Sci. Technol.* **2008**, *42*, 5675–5681. [CrossRef] [PubMed]

13. Marcelli, A.; Cibin, G.; Hampai, D.; Giannone, F.; Sala, M.; Pignotti, S.; Maggi, V.; Marino, F. XANES characterization of deep ice core insoluble dust in the ppb range. *J. Anal. At. Spectrom.* **2012**, *27*, 33–37. [CrossRef]

14. Marcelli, A.; Maggi, V. *Aerosols in Snow and Ice. Markers of Environmental Pollution and Climatic Changes: European and Asian Perspectives*; Superstripes Press: Rome, Italy, 2017; ISBN 9788866830771.

15. D'Elia, A.; Cibin, G.; Robbins, P.E.; Maggi, V.; Marcelli, A. Design and characterization of a mapping device optimized to collect XRD patterns from highly inhomogeneous and low density powder samples. *Nucl. Instrum. Methods Phys. Res. B* **2017**, *411*, 22–28. [CrossRef]

Article

Iron Speciation in Insoluble Dust from High-Latitude Snow: An X-ray Absorption Spectroscopy Study

Shiwei Liu [1,2], Cunde Xiao [3], Zhiheng Du [1,*], Augusto Marcelli [4,5,*], Giannantonio Cibin [6], Giovanni Baccolo [7,8], Yingcai Zhu [9], Alessandro Puri [10], Valter Maggi [7,8] and Wei Xu [9]

1   State Key Laboratory of Cryospheric Science, Northwest Institute of Eco-Environment and Resources, Chinese Academy of Sciences, Lanzhou 730000, China; liushiwei1990@lzb.ac.cn
2   University of Chinese Academy of Science, Beijing 100049, China
3   State Key Laboratory of Land Surface Processes and Resource Ecology, Beijing Normal University, 19 Xinjiekouwai Street, Beijing, 100875, China; cdxiao@bnu.edu.cn
4   Istituto Nazionale di Fisica Nucleare, Laboratori Nazionali di Frascati, 00044 Frascati, Italy
5   RICMASS, Rome International Center for Materials Science Superstripes, 00185 Rome, Italy
6   Diamond Light Source, Harwell Science and Innovation Campus, Didcot OX11 0DE, UK; giannantonio.cibin@diamond.ac.uk
7   Earth and Environmental Sciences Department, University Milano-Bicocca, 20126 Milano, Italy; giovanni.baccolo@mib.infn.it (G.B.); valter.maggi@unimib.it (V.M.)
8   INFN-Milan Bicocca Section, 20126 Milan, Italy
9   Beijing Synchrotron Radiation Facility, Institute of High Energy Physics, Chinese Academy of Sciences, Beijing,100049, China; yingcaizhu@ihep.ac.cn (Y.Z.); xuw@mail.ihep.ac.cn (W.X.)
10 CNR-IOM-OGG, c/o ESRF, 38043 Grenoble, France; alessandro.puri@esrf.fr
*   Correspondence: duzhiheng10@163.com (Z.D.); marcelli@lnf.infn.it (A.M.)

Received: 1 October 2018; Accepted: 3 December 2018; Published: 10 December 2018

Abstract: Iron is thought to limit the biomass of phytoplankton populations in extensive regions of the ocean, which are referred to as high-nutrient low-chlorophyll (HNLC) regions. Iron speciation in soils is still poorly understood. We have investigated inorganic and organic standard substances, diluted mixtures of common Fe minerals in insoluble dust in snow from the Laohugou No.12 glacier, and sand (including soil and moraine) samples that were collected from western China. The speciation of iron (Fe) in insoluble dust and sand was determined by X-ray absorption near-edge structure (XANES) spectroscopy. A linear fit combination (LCF) analysis of the experimental spectra compared to a large set of reference compounds showed that all spectra can be fitted by only four species: Fe_2O_3, Fe_3O_4, biotite, and ferrous oxalate dihydrate (FOD). A significant altitude effect was detected for snow. The proportion of Fe_2O_3 in snow decreases gradually, and vice versa for FOD. As for Fe_3O_4 and biotite, the altitude effect was also detected, but separate regions should be considered to be deduced by topography. The Fe species in moraines and soils were also analyzed to identify the source of moraines and the heterogeneity of soils, and were compared with snow.

Keywords: Laohugou glacier; snow; insoluble dust; iron speciation; XANES and LCF

---

1. Introduction

Iron (Fe) contributes 5.1 mass percent to the earth's crust, and is a major component of many soil-forming parent materials. Consequently, primary or secondary Fe-containing minerals, such as olivine, biotite, pyrite, ferrihydrite, goethite, haematite, lepidocrocite, and Fe-bearing clay minerals, are significant components in most soils [1]. The presence or absence of pedogenic Fe minerals in soils, soil horizons, or soil aggregates, as well as their spatial distribution within a soil profile or an aggregate, is strongly related to the ambient physicochemical conditions, such as pH, redox potential, and activity of organic ligands [2,3]. At variance, soil physicochemical conditions are greatly influenced by

Fe-containing minerals, because they can undergo redox, sorption, (de)protonation, or (co)precipitation reactions with organic and inorganic constituents of the soil solution and the soil solid phase [4]. Because most Fe minerals are strongly colored and their presence or absence is clearly visible in soil profiles [5,6], they have been widely used to identify pedogenic processes (podzolisation, gleysation, lessivage) and to classify soils [2,4].

As a consequence of the considerable importance of Fe-containing minerals in the earth sciences, their identification and quantification in geologic materials and soils has been a major target of research, and numerous methods for the assessment of Fe minerals in soils have been developed. These methods include X-ray diffraction [7], wet chemical fractionation procedures, such as selective extraction [8–10], differential thermal analysis [11], Mössbauer spectroscopy [7], and colorimetry [5,6]. However, most of these methods are either applicable only to crystalline phases (XRD) or provide only operationally defined results (dissolution methods). Moreover, none of these methods provides information on the micro-morphological arrangement of different Fe species in natural soil structures, e.g., aggregates.

Aerosols originating from soils of arid regions are probably the main source of bioavailable Fe in the open ocean. However, recently, a more complex scenario, with multiple environmental sources for aerosol containing iron, is emerging [12]. Although aerosol solubility could be heavily affected by Fe speciation, which in turn could vary considerably among possible aerosol source environments, no such relationship has yet been clearly established, largely owing to a lack of direct measurements of particle speciation. Synchrotron radiation (SR) is an ideal X-ray source that provides a high intensity photon for the investigation of Fe speciation.

X-ray fluorescence (XRF) and X-ray absorption near-edge structure (XANES) spectroscopy are major SR techniques. XANES spectroscopy provides information on the valence state and binding structure of absorbing elements. XANES was first shown to be determined by multiple scattering resonances of the photoelectron that is emitted at the absorbing atomic site by Belli et al. [13], who probed the local structure of a nanoscale cluster of atoms centered at the absorbing atomic site. Bianconi et al. [14] used XANES spectroscopy to quantitatively probe into subtle local lattice distortions of the Fe site. XANES spectroscopy is widely used as a tool for local structure investigations [15], and has been used for chemical state analyses on a wide range of Fe elements [16–19].

A linear combination fit (LCF) analysis has previously been chosen to directly determine the substance speciation in soil, aerosol, and sediment [19–22]. This technique has already been applied to the investigation of the local molecular structure in the region of selected metal atoms in homogenous dust molecules. In the present study, XANES and LCF were applied to investigate the Fe speciation in dust dispersed in snow on the LHG glacier, which is located in western China. Several local and foreign soils were also measured in an attempt to identify the dust's source.

2. Materials and Methods

2.1. Sample Collection

The Laohugou (LHG) NO.12 glacier (5Y448D0012) is one of the largest valley glacier groups on Qilian Mountain, which is located at the northern edge of the Tibetan Plateau adjacent to the Gobi desert. The LHG No.12 glacier is a typical continental glacier [23] that is located on the north slope of western Qilian Mountain. It covers an area of 21.9 km^2 with a large accumulation zone. Its terminal elevation is 4260 m above sea level (a.s.l.). Above 4500 m a.s.l., the glacier is flat, while below 4500 m a.s.l., a large serac area is present.

Fresh surface snow samples were collected at the site shown in Figure 1 with a 500 mL LDPE bottle at different altitudes (4300 m, 4400 m, 4500 m, 4600 m, 4700 m, and 4900 m) on 25 May 2016. In the laboratory, the collected snow samples were melted rapidly at room temperature (~25 °C). After melting, all samples were filtered immediately to minimize the adhesion of particles to the surface of the bag by pouring the snow-sourced water into a 500 mL percolator and pumping the sample slowly through an oven-dried (at 60 °C for 24 h), 70 μm plastic filter. After, the filters were removed from the holders and placed into plastic captures, which held them in place under a dry and low temperature condition. The plastic filters with dust were used for the XRF experiments. Besides this, one soil sample from the end of the LHG glacier (4200 m a.s.l.) and two moraine samples from the LHG tongue at ~4250 and 4700 m a.s.l. were excavated. In order to compare the snow samples with LHG soil, and trace the source of the insoluble content in snow, several soil and sand samples around the LHG region were also collected. Table 1 summarizes the information on the samples. The moraine, soil, and sand samples were dried at room temperature (~25 °C) for a few days, then powdered and filtered by a 70 μm agate filter. We then prepared pellets of ~2 g each for the X-ray absorption measurement.

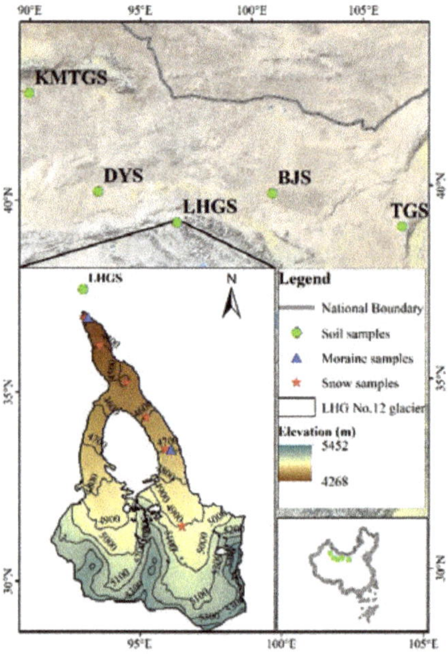

Figure 1. The location of the Laohugou (LHG) glacier and the collection sites of soils and sands described in Table 1.

Table 1. The collected samples.

Sample Name	Substance Configuration	Collecting Location
LHGS	soil	End of the LHG No.12 glacier
KMTGS	sand	Kumtag desert
DYS	soil	Dunhuang yadan
TGS	sand	Tengger desert
BJS	sand	Badain Jaran desert
Tongue moraine	moraine	4250 m a.s.l. LHG No.12 glacier
4700 m moraine	moraine	4700 m a.s.l. LHG No.12 glacier
4300 m snow	insoluble dust in snow	4300 m a.s.l. LHG No.12 glacier
4400 m snow	insoluble dust in snow	4400 m a.s.l. LHG No.12 glacier
4500 m snow	insoluble dust in snow	4500 m a.s.l. LHG No.12 glacier
4600 m snow	insoluble dust in snow	4600 m a.s.l. LHG No.12 glacier
4700 m snow	insoluble dust in snow	4700 m a.s.l. LHG No.12 glacier
4900 m snow	insoluble dust in snow	4900 m a.s.l. LHG No.12 glacier

2.2. Choice of Reference Compounds

The X-ray Fluorescence (XRF) lines of each element are defined by the transition energy among different orbitals. Each line occurs at a fixed energy and the fluorescence lines are clearly separated for different elements. The energy value of each emission line allows us to identify the element in the investigated sample. For instance, the K_α and K_β lines of iron can be detected at 6.4 keV and 7.06 keV, respectively. The L_α, L_β, and L_γ lines of lead occur at 10.6 keV, 12.6 keV, and 14.8 keV, respectively. Usually, one should at least rely on two lines to justify the existence of the corresponding element because, due to a poor energy resolution, the XRF lines of some elements may overlap and cannot be distinguished. For instance, the K_α line of arsenic is quite close to the L_α line of lead.

2.3. X-ray Absorption Measurements

The iron speciation in the samples was determined using synchrotron-based techniques, such as X-ray Absorption Near-Edge Structure (XANES) spectroscopy and microscopic X-ray fluorescence measurements at LISA (Linea Italiana per la Spettroscopia di Assorbimento di raggi X), the BM08 beamline of the European Synchrotron Radiation Facility (ESRF). The energy has been selected with a double crystal monochromator using a Si(111) crystal pair and two Si-coated mirrors for harmonics rejection. Further details about the beamline and the technical specifications are available elsewhere [24,25]. We worked at room temperature, and, due to the flux available on the bending magnet beamline, we never observed radiation damage to our samples.

The preliminary Fe K-edge and XRF data results indicate that they can be used to trace the variability of mineralogy on snow, ice, and solid samples with different provenances. Energy was calibrated by setting to 7112 eV the energy position of the first inflection point in the K-edge of a Fe foil recorded in a double transmission setup. For each snow sample and solid sample, one to four scans were recorded, depending on the sample content and Fe concentration. Spectra have been collected under vacuum (10–2 mbar) at ambient temperature in fluorescence mode. We used a four-channel SDD (Solid State Detector) detector with an energy resolution of ~180 eV at the Fe $k\alpha$ emission line (~6400 eV). In the case of filters, due to the large dimensions and the inhomogeneity, an unfocused beam of ~2 mm^2 was used to probe these samples. Data were averaged if necessary and normalized using the Athena [26] software. EXAFS spectra were then background subtracted from these normalized data using the XAFS code [27].

3. Results and Discussions

3.1. Fe K-Edge XANES of Standards

Ten iron standard natural compounds, one mineral, and one organic iron system were selected as references. Figure 2 shows the XANES spectra at the iron K-edge of the 12 standard substances listed in Table 1. The standard substances in which iron is in the Fe (II) state include $FeSO_4$, $FeCl_2$, FeS_2, and FeS, and those in which iron is in the Fe (III) state are $FeCl_3$, $FePO_4$, Fe_2O_3, $Fe(SO_4)_3$, and $Fe(NO_4)_3$, including the Fe_3O_4. These standards showed clear differences in the energy positions of the centroids of their normalized pre-edge peaks, of the energy position, and of their white lines (Table 1). Typically, the Fe absorption edge shifts upward from 7120.5 to 7128.6, increasing the oxidation state. The edge position of Fe (II) and Fe (III) compounds falls in the ranges 7116–7121 eV and 7126–7129 eV, respectively. According to Figure 2b, an intense pre-edge peak is detected in FeS_2, FeS, Fe_3O_4, and Fe_2O_3. The pre-edge position for Fe_3O_4 occurs at a lower energy than that for Fe_2O_3, biotite, and ferrous oxalate dihydrate (FOD).

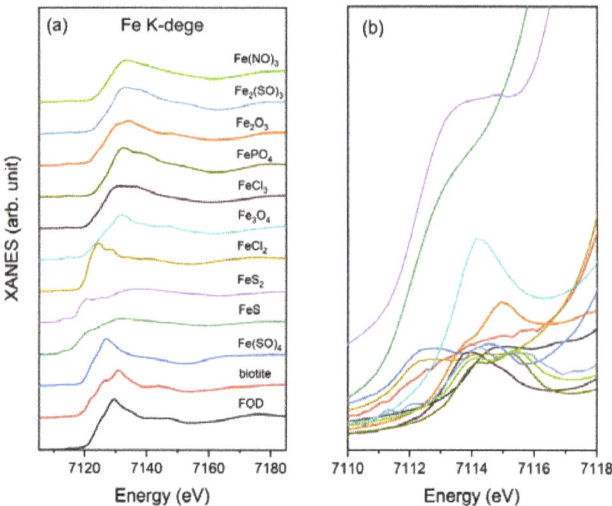

Figure 2. (**a**) A comparison of the XANES spectra of iron standards (**b**) and the pre-edge region of the iron K-edge of the spectra shown in the left panel.

3.2. Fe K-Edge XANES of the Soil, Moraine, and Snow Samples

Fe speciation is of great interest due to its role in the dissolution process of Fe from dust minerals and for the possibility to probe into the contribution of soluble iron to the ocean. Indeed, Fe is correlated to phytoplankton growth in the sea and affects the oceanic biogeochemical cycle [28,29]. It is, thus, fundamental to probing and understanding the different factors that affect Fe speciation at a regional scale, to further constrain the role of iron in biogeochemical models [30].

The Fe-edge XANES spectra of the soil of the LHG glacier (LHGS sample) and the moraine samples are shown in Figure 3a, and those of the other relevant soil samples and of the six snow samples are shown in Figure 3b,c, respectively. The result indicates that biotite and Fe_2O_3 are the main components in these samples. Although the Fe speciation in biotite is Fe^{2+}, the Fe oxide (Fe_2O_3) in these samples indicates that Fe^{3+} is the primary component because Fe^{3+} is much more stable. For comparison, the spectra of two analogous reference compounds, i.e., Fe(II)-biotite and Fe(III)-Fe_2O_3, have been included in all figures. Moraine is the sediment that is transported by the glacier accumulation process, which mainly comes from a fragment of a mountain. The Fe-edge XANES spectra of the moraines and LHGS are similar and closely resemble biotite, indicating that the iron in both the LHGS and moraine is primarily associated with biotite. This provides another explanation for the moraine source on the glacier, although at the molecule level. Figure 3b compares the Fe-edge XANES spectra of relevant soils and sands surrounding the LHG glacier. The spectra of soil and sand samples resemble those of biotite and Fe_2O_3, but are a complex mixture of iron minerals, and not just a mixture of biotite and Fe_2O_3.

The result makes it possible to distinguish the LHGS sample. Figure 3c shows the Fe-edge XANES spectra of the snow samples and relevant molecules of the iron element. The spectra of the snow samples are homologous but inconsistent with the spectra of biotite and Fe_2O_3, pointing out that the insoluble dust of the snow samples may not come only from a local dust source (i.e., the LHGS or moraine), but may contain other reference compounds.

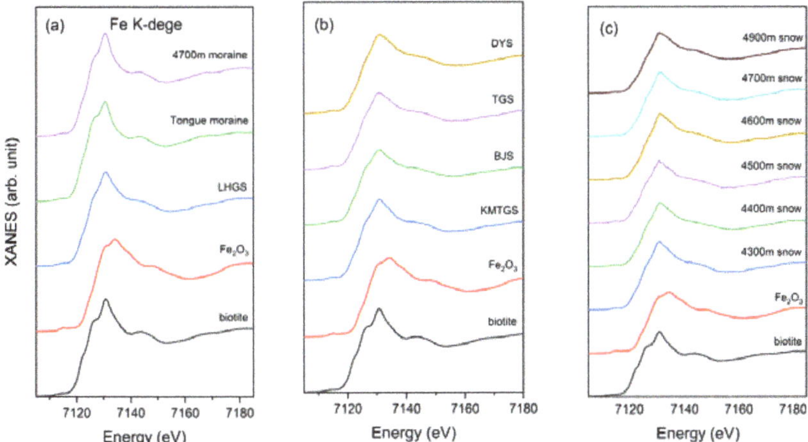

Figure 3. XANES spectra at the iron K-edge of (**a**) the soil and moraine samples; (**b**) the reference soil; and (**c**) the snow samples.

3.3. Linear Combination Fitting

The k^2-weighted Fe K-edge XANES spectra of all samples were analyzed by a principal component analysis (PCA) [20] to determine the number of reference model spectra needed to simulate the experimental data. Based on the PCA results, a target transformation (TT) was further performed in order to select the reference standard substances that were most likely present in these samples [31]. After that, based on the results from the PCA/TT, a linear combination fitting (LCF) analysis was performed using the software Athena to calculate the proportion of each iron reference in the soil, moraine, and snow samples. Several standard substances were selected to represent the possible iron compounds that were potentially present in the LHG samples. The PCA/TT result shows that Fe_2O_3, Fe_3O_4, biotite, and ferrous oxalate dihydrate (FOD) are the major standard substances in the soil samples, moraine samples, and snow samples. The four reference compounds were used to run the LCF analysis. A fit range of -20 to 50 eV was selected to fit the sample spectra. The LCF results are summarized in Table 2.

Table 2. The linear combination fitting (LCF) results of the soil, moraine, and snow samples. FOD, is the ferrous oxalate dihydrate.

Samples	Fe_2O_3	Fe_3O_4	Biotite	FOD	R-Factor
	Proportion (Uncertainty)				
LHGS	0.080 (0.025)	0.218 (0.019)	0.524 (0.013)	0.178 (0.008)	0.0002
Tongue moraine	/	/	0.738 (0.089)	0.262 (0.102)	0.0017
4700 m moraine	/	/	0.719 (0.018)	0.281 (0.018)	0.0022
KMTGS	0.307 (0.061)	0.354 (0.038)	0.196 (0.034)	0.143 (0.018)	0.0010
DYS	0.346 (0.031)	0.235 (0.026)	0.338 (0.014)	0.081 (0.008)	0.0003
TGS	0.192 (0.018)	0.197 (0.015)	0.417 (0.010)	0.194 (0.005)	0.0002
BJS	0.151 (0.035)	0.147 (0.026)	0.514 (0.019)	0.188 (0.009)	0.0005
4300 m snow	0.478 (0.094)	0.208 (0.148)	/	0.315 (0.050)	0.0075
4400 m snow	0.264 (0.058)	0.197 (0.043)	0.275 (0.027)	0.264 (0.016)	0.0006
4500 m snow	0.248 (0.053)	0.092 (0.042)	0.479 (0.026)	0.101 (0.019)	0.0000
4600 m snow	0.200 (0.114)	0.479 (0.071)	/	0.320 (0.036)	0.0040
4700 m snow	0.194 (0.036)	0.187 (0.099)	0.278 (0.028)	0.341 (0.017)	0.0009
4900 m snow	0.168 (0.048)	/	0.468 (0.027)	0.365 (0.034)	0.0044

The spectra of all soil, moraine, and snow samples closely resemble a combination of Fe_2O_3, Fe_3O_4, biotite, and FOD. The spectra of the LHG moraines are closer to biotite and FOD, and the overall curvature of the Fe-XANES spectrum for the moraines is similar to that for biotite, which demonstrates that the two reference compounds are the components of moraine. As shown in Figure 4, the LCF result shows that biotite is the major component of both moraines with a proportion of 73.8% and 71.9% in the LHG tongue and 4700 a.s.l. samples, respectively. At the same time, the minor component is FOD with a proportion of 26.2% and 28.1%, respectively. The two moraine samples come from mountain soil and debris, which accumulate by the glacier's movement.

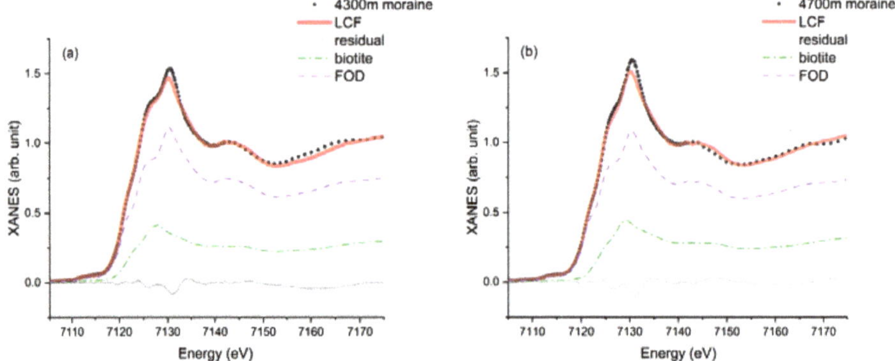

Figure 4. Iron speciation in the moraine samples, as calculated by the linear combination fitting (LCF) of Fe K-edge XANES spectra of two moraine samples collected at 4300 m (**a**) and at 4700 m (**b**).

There is a different fraction of reference compounds between the LHGS and LHG moraines, which means that it is difficult to be sure that all the moraines come from local soil. The LHGS sample was collected at the end of the LHG No.12 glacier, which is 4200 m a.s.l.. The LCF analysis points out that biotite is the major component, with 52.4% of reference compounds. The difference among our soil samples and moraines could be due to the heterogeneity from the mountain soils. However, spectral features and the LCF analysis point out that the LHGS sample and the moraines have a similar iron composition, which provides another explanation for the moraine source on the glacier, although at the molecule level. The Fe speciation can distinguish local soil samples (LHGS) from potential desert sources (i.e., KMTGS, DYS, TGS, and BJS) using the LCF method (Table 3). The LCF spectra of the LHGS and KMTGS samples are shown in Figure 5. The other samples were also fitted, but are not shown.

Table 3. The energy positions of the pre-edge peak centroid, the inflection point of the absorption edge, and the white line of the Fe K-edge XANES spectra acquired for different Fe-bearing standards.

Compound	Fe Oxidation State	Centroid Pre-Edge Peak	Inflection Point Absorption Edge	White Line
			Energy Position/eV	
biotite	+2	7114.9	7121.0	7127.9
FOD	+2	7115.1	7121.0	7127.9
$FeSO_4$	+2	7114.6	7120.5	7127.0
FeS	+2	7113.2	7116.4	7131.8
FeS_2	+2	7114.8	7116.6	7120.8
$FeCl_2$	+2	7114.2	7119.4	7124.5
Fe_3O_4	+2/+3	7114.2	7128.3	7132.0
$FeCl_3$	+3	7114.1	7126.3	7131.9
$FePO_4$	+3	7115.2	7127.2	7132.6
Fe_2O_3	+3	7114.9	7125.9	7134.1
$Fe(SO_4)_3$	+3	7114.2	7128.6	7133.3
$Fe(NO4)_3$	+3	7115.2	7127.3	7133.5

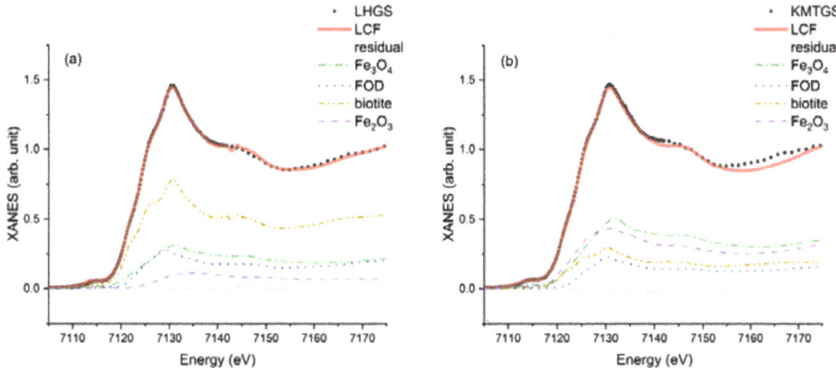

Figure 5. The iron contribution to the desert samples that were collected at different regions, as calculated by linear combination fitting (LCF) from the Fe K-edge XANES spectra of the LHG No.12 glacier (**a**) and the Kumtag desert (**b**).

The fraction of biotite in the LHGS and in the BJS is comparatively the same, more than 50%, which is the highest fraction, and is followed by the TGS, DYS, and KMTG with a proportion of 41.7%, 33.8%, and 19.6%, respectively. The ratio of Fe_2O_3 to Fe_3O_4 in the endemic soils is almost equal to one. However, the proportion of Fe_2O_3 in the LHGS is 8%, whereas the proportion of Fe_3O_4 is 21.8%. That makes it possible to separate the LHG from the ecdemic soils within the local source and the far source contributions of the insoluble dust to the snow precipitation. For the FOD, almost all of the soil samples are similar, except for DYS. They contain less FOD than the LHG moraines, probably due to the dry conditions and the different soil texture. The proportion of reference compounds in all measured soil (sand) samples is shown in Figure 6. A slow change in the proportion is observed and makes it possible to identify different sources of the insoluble dust in the LHG snow samples.

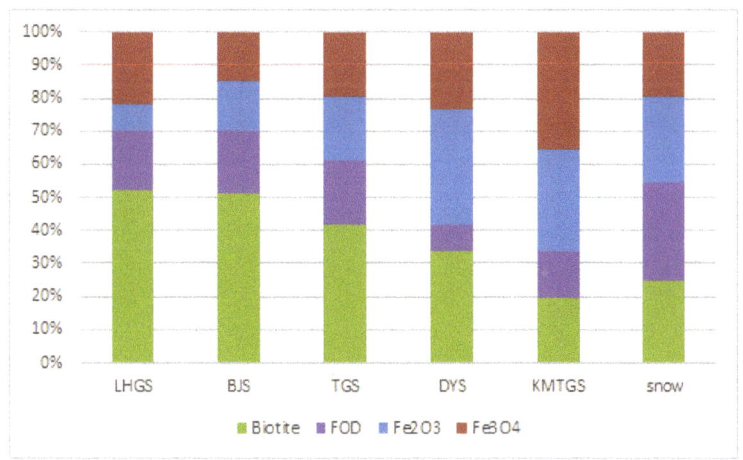

Figure 6. The different proportions of the four iron components in both the soil and snow samples showed in Figure 5.

Figure 7 shows the Fe K-edge XANES spectra of snow samples from different altitudes (4400 m and 4700 m a.s.l. are shown in (a) and (b), respectively) in the LHG glacier together with the Fe reference compounds that yield the best fits by the LCF analysis. The R-factor of LCF indicates that the fit is acceptable, and half of the six snow samples are composed of four iron components, except for the 4300 m, 4600 m, and 4900 m snow samples, for which the LCF performed better with only three reference components.

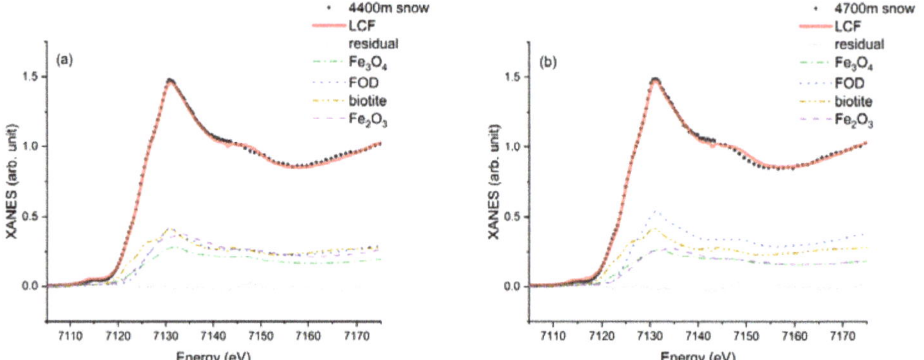

Figure 7. Iron speciation in different altitude snow samples, as calculated by linear combination fitting (LCF) from the Fe K-edge XANES spectra of the snow samples collected at 4400 m (**a**) and at 4700 m (**b**).

The proportion of iron components in the snow samples is shown in Figure 8. Fe_2O_3, Fe_3O_4, biotite, and FOD share a comparable proportion in the average of all snow samples, with the average percentage of 25.9%, 19.4%, 25.0%, and 29.8%, respectively. According to Figure 8, these snow samples can be separated into two sets of altitudes (i.e., 4300–4500 m and 4600–4900 m). In both categories, the snow samples contain more biotite and less Fe_2O_3 and Fe_3O_4. At a lower altitude, biotite is not a reference component in the 4300 m snow sample, and the content of Fe_2O_3, Fe_3O_4, and FOD is 47.8%, 25.8%, and 31.5%, respectively. With an increase in the elevation, biotite becomes a major component (up to 47.9%), and the proportion of the other reference components decreases.

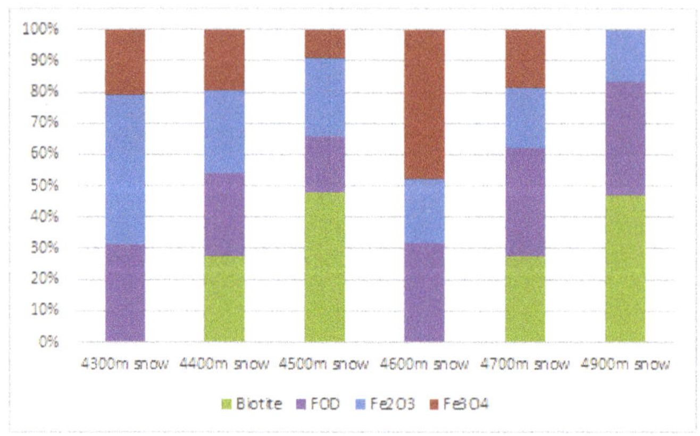

Figure 8. The proportion of the four iron components in the snow samples collected at different altitudes.

The average proportion of Fe_2O_3, Fe_3O_4, biotite, and FOD at a lower altitude by LCF is 33%, 16.6%, 25.1%, and 25.3%, respectively. For the higher altitude region, biotite is also not a component. In the lowest altitude snow sample, Fe_3O_4 is the major component. With an increase in the elevation, biotite becomes the major component (up to 36.5%), and the proportion of the other reference components is reduced. Moreover, Fe_3O_4 disappears at 4900 m. The average proportion of Fe_2O_3,

Fe_3O_4, biotite, and FOD in the lower altitude snow samples, as calculated by LCF, is 18.8%, 22.2%, 24.9%, and 34.2%, respectively.

When all of the regular patterns of reference components in the snow samples collected at different altitudes are combined, a clear effect versus altitude emerges. The proportion of Fe_2O_3 in the snow samples decreases gradually. The proportion of FOD in the snow samples clearly decreases at low altitude (4300–4550 m a.s.l.) and increases weakly at high altitude (4600–4900 m a.s.l.). As for Fe_3O_4 and biotite, an altitude effect could also be detected; however, separate regions should be considered by topography. The absence of biotite and Fe_3O_4 in several snow samples makes it difficult to confirm the source of precipitation and the mechanism that is associated with the altitude. A deeper investigation is in progress, and additional soil and snow samples will be investigated.

4. Conclusions

This study investigates the speciation of Fe in insoluble dust and sand samples from western China by means of X-ray absorption spectroscopy techniques and in particular by XANES spectroscopy. The insoluble dust in snow and sand samples represents the contribution of typical arid regions. The linear combination fit of experimental XANES spectra of selected relevant Fe compounds suggests that Fe in the insoluble dust is mainly present as inorganic species, e.g., Fe_2O_3, Fe_3O_4, biotite, and ferrous oxalate dihydrate. Biotite is the major component in the LHG soil and moraines, and appears also to be a significant component in the snow samples. There is an altitude effect of the reference components on the insoluble dust in snow: the proportion of Fe_2O_3 in the snow decreases gradually, and the proportion of FOD in the snow clearly decreases at low altitude (4300–4550 m a.s.l.) and increases weakly at high altitude (4600–4900 m a.s.l.). As for Fe_3O_4 and biotite, the altitude effect could also be detected, but a better analysis of the investigated region should be considered as being deduced by topography.

The spectra of the LHG moraines are closer to biotite and FOD, and the overall shape of the XANES spectra at the Fe K-edge for moraines is similar to the biotite spectrum. The results demonstrate that the two reference compounds are the major components of moraine, and that the LHGS and moraines have a similar iron composition. The data provide an alternative explanation for the moraine source in the glacier for different Fe contents and valences. A gradual change in the composition is evident in the soil samples from local and long-distance sources. However, additional work on Fe composition and speciation in other samples is necessary to better understand the mineral dust transport mechanisms.

Author Contributions: S.L., C.X., Z.D., A.M., and W.X. designed the study. Z.D. and S.L. performed the field work. S.L., Z.D., A.M., G.C., A.P., G.B., and Y.Z. performed the experiments at ESRF. S.L., Y.Z., Z.D., and W.X. performed the XAS data analysis. All authors wrote the manuscript and contributed to the discussion/interpretation of the results.

Funding: This research was funded by Cunde Xiao grant number 41425003, Zhiheng Du grant number 41701071, Wei Xu grant number U1532128, and the CAS "Light of West China" Program.

Acknowledgments: We strongly acknowledge the staff of the Italian CRG LISA for their support at ESRF within the experiment 08-01-1031 on Beamline BM08 and for many fruitful discussions. The support of DARA Department of the Italian *Presidenza del Consiglio dei Ministri* is gratefully acknowledged.

Conflicts of Interest: No conflicts of interest.

References

1. Allen, B.L.; Hajek, B.F. Mineral occurrence in soil environments. *Miner. Soil Environ.* **1989**, *28*, 1–52.
2. Blume, H.P.; Schwertmann, U. Genetic evaluation of profile distribution of aluminum, iron, and manganese oxides 1. *Soil Sci. Soc. Am. J.* **1969**, *33*, 438–444. [CrossRef]
3. Cornell, R.M. The iron oxides. In *Soil Mineralogy with Environmental Applications*; Soil Science Society of America: Madison, WI, USA, 1996; pp. 363–369.
4. Schwertmann, U. Relations between iron oxides, soil color, and soil formation. *Soil Color* **1993**, *31*, 51–69.

5.   Schwertmann, U.; Schwertmann, U.; Lentze, W. Bodenfarbe und eisenoxidform. *Z. Pflanzenernährung Düngung Bodenkd.* **1966**, *115*, 209–214. [CrossRef]

6.   Scheinost, A.; Schwertmann, U. Color identification of iron oxides and hydroxysulfatesuse and limitations. *Soil Sci. Soc. Am. J.* **1999**, *5*, 1463–1471.

7.   Schwertmann, U.; Fechter, H. The point of zero charge of natural and synthetic ferrihydrites and its relation to adsorbed silicate. *Clay Miner.* **1982**, *17*, 471–476. [CrossRef]

8.   Mehra, O.; Jackson, M. Iron oxide removal from soils and clays by a dithionite–citrate system buffered with sodium bicarbonate. In Proceedings of the Seventh National Conference on Clays and Clay Minerals; Elsevier: Amsterdam, The Netherlands, 1960; pp. 317–327.

9.   Schwertmann, U. Differenzierung der eisenoxide des bodens durch extraktion mit ammoniumoxalat-lösung. *Z. Pflanzenernährung Düngung Bodenkd.* **1964**, *105*, 194–202. [CrossRef]

10. Holmgren, G.G. A rapid citrate-dithionite extractable iron procedure 1. *Soil Sci. Soc. Am. J.* **1967**, *31*, 210–211. [CrossRef]

11. Vold, M.J. Differential thermal analysis. *Anal. Chem.* **1949**, *21*, 683–688. [CrossRef]

12. Li, W.; Xu, L.; Liu, X.; Zhang, J.; Lin, Y.; Yao, X.; Gao, H.; Zhang, D.; Chen, J.; Wang, W. Air pollution–aerosol interactions produce more bioavailable iron for ocean ecosystems. *Sci. Adv.* **2017**, *3*, e1601749. [CrossRef]

13. Belli, M.; Scafati, A.; Bianconi, A.; Mobilio, S.; Palladino, L.; Reale, A.; Burattini, E. X-ray absorption near edge structures (xanes) in simple and complex mn compounds. *Solid State Commun.* **1980**, *35*, 355–361. [CrossRef]

14. Bianconi, A.; Dell'Ariccia, M.; Durham, P.J.; Pendry, J.B. Multiple-scattering resonances and structural effects in the x-ray-absorption near-edge spectra of fe ii and fe iii hexacyanide complexes. *Phys. Rev. B* **1982**, *26*, 159–167. [CrossRef]

15. Bachrach, R.Z. *Synchrotron Radiation Research: Advances in Surface and Interface Science*; Kluwer Academic: Dordrecht, The Netherlands, 1992.

16. Kwiatek, W.; Gałka, M.; Hanson, A.; Paluszkiewicz, C.; Cichocki, T. Xanes as a tool for iron oxidation state determination in tissues. *J. Alloys Compd.* **2001**, *328*, 276–282. [CrossRef]

17. Bajt, S.; Sutton, S.; Delaney, J. X-ray microprobe analysis of iron oxidation states in silicates and oxides using x-ray absorption near edge structure (xanes). *Geochim. Cosmochim. Acta* **1994**, *58*, 5209–5214. [CrossRef]

18. Galoisy, L.; Calas, G.; Arrio, M. High-resolution xanes spectra of iron in minerals and glasses: Structural information from the pre-edge region. *Chem. Geol.* **2001**, *174*, 307–319. [CrossRef]

19. Prietzel, J.; Thieme, J.; Eusterhues, K.; Eichert, D. Iron speciation in soils and soil aggregates by synchrotron-based x-ray microspectroscopy (xanes, μ-xanes). *Eur. J. Soil Sci.* **2007**, *58*, 1027–1041. [CrossRef]

20. Ressler, T.; Wong, J.; Joseph Roos, A.; Smith§, I.L. Quantitative speciation of mn-bearing particulates emitted from autos burning (methylcyclopentadienyl)manganese tricarbonyl-added gasolines using xanes spectroscopy. *Environ. Sci. Technol.* **2000**, *34*, 950–958. [CrossRef]

21. Takahashi, Y.; Higashi, M.; Furukawa, T.; Mitsunobu, S. Change of iron species and iron solubility in asian dust during the long-range transport from western china to japan. *Atmos. Chem. Phys.* **2011**, *11*, 11237–11252. [CrossRef]

22. Kraal, P.; Burton, E.D.; Rose, A.L.; Kocar, B.D.; Lockhart, R.S.; Grice, K.; Bush, R.T.; Tan, E.; Webb, S.M. Sedimentary iron–phosphorus cycling under contrasting redox conditions in a eutrophic estuary. *Chem. Geol.* **2015**, *392*, 19–31. [CrossRef]

23. Shi, Y. *Concise Glacier Inventory of China*; Shanghai Popular Science Press: Shanghai, China, 2008.

24. d'Acapito, F.; Trapananti, A.; Puri, A. Lisa: The italian crg beamline for x-ray absorption spectroscopy at esrf. *J. Phys. Conf. Ser.* **2016**, *712*, 012021. [CrossRef]

25. d'Acapito, F. *Lisa Annual Report 2017*; CNR-IOM: Grenoble, France, 2017.

26. Ravel, B.; Newville, M. Athena, artemis, hephaestus: Data analysis for x-ray absorption spectroscopy using ifeffit. *J. Synchrotron Radiat.* **2010**, *12*, 537–541. [CrossRef] [PubMed]

27. Winterer, M. Xafs—A data analysis program for materials science. *J. Phys. IV* **1997**, *7*, C2-243–C2-244. [CrossRef]

28. Coale, K.H.; Johnson, K.S.; Fitzwater, S.E.; Gordon, R.M.; Tanner, S.; Chavez, F.P.; Ferioli, L.; Sakamoto, C.; Rogers, P.; Millero, F. A massive phytoplankton bloom induced by an ecosystem-scale iron fertilization experiment in the equatorial pacific ocean. *Nature* **1996**, *383*, 495–501. [CrossRef] [PubMed]

29. Jickells, T.D.; An, Z.S.; Andersen, K.K.; Baker, A.R.; Bergametti, G.; Brooks, N.; Cao, J.J.; Boyd, P.W.; Duce, R.A.; Hunter, K.A.; et al. Global iron connections between desert dust, ocean biogeochemistry, and climate. *Science* **2005**, *308*, 67–71. [CrossRef] [PubMed]

30. Mahowald, N.M.; Engelstaedter, S.; Luo, C.; Sealy, A.; Artaxo, P.; Beniteznelson, C.; Bonnet, S.; Chen, Y.; Chuang, P.Y.; Cohen, D.D. Atmospheric iron deposition: Global distribution, variability, and human perturbations. *Ann. Rev. Mar. Sci.* **2009**, *1*, 245–278. [CrossRef] [PubMed]

31. Noël, V.; Marchand, C.; Juillot, F.; Ona-Nguema, G.; Viollier, E.; Marakovic, G.; Olivi, L.; Delbes, L.; Gelebart, F.; Morin, G. Exafs analysis of iron cycling in mangrove sediments downstream a lateritized ultramafic watershed (vavouto bay, new caledonia). *Geochim. Cosmochim. Acta* **2014**, *136*, 211–228. [CrossRef]

Article

XANES Iron Geochemistry in the Mineral Dust of the Talos Dome Ice Core (Antarctica) and the Southern Hemisphere Potential Source Areas

Valter Maggi [1,2,3,*], Giovanni Baccolo [1,2,4], Giannantonio Cibin [5], Barbara Delmonte [1], Dariush Hampai [6] and Augusto Marcelli [6,7]

[1] Department of Earth and Environmental Sciences, University of Milano-Bicocca, 20126 Milano, Italy; giovanni.baccolo@mib.infn.it (G.B.); barbara.delmonte@unimib.it (B.D.)
[2] INFN Section of Milano-Bicocca, 20126 Milano, Italy
[3] IGG-CNR, 56100 Pisa, Italy
[4] Graduate school in Polar Sciences, University of Siena, 53100 Siena, Italy
[5] Diamond Light Source, Didcot OX11 0AB, UK; giannantonio.cibin@diamond.ac.uk
[6] INFN-Laboratori Nazionali di Frascati, Via Enrico Fermi 40, 00044 Frascati, Italy; dariush.hampai@lnf.infn.it (D.H.); augusto.marcelli@lnf.infn.it (A.M.)
[7] RICMASS, Rome International Center for Materials Science Superstripes, Via dei Sabelli 119A, 00185 Rome, Italy
* Correspondence: valter.maggi@unimib.it; Tel.: +39-026-448-2874

Received: 27 June 2018; Accepted: 22 November 2018; Published: 6 December 2018

Abstract: X-ray absorption near edge structure (XANES) measurements at the Fe K-edge were performed on aeolian dust in the TALos Dome Ice CorE drilling project (TALDICE) ice core drilled in the peripheral East Antarctic plateau, as well as on Southern Hemisphere potential source area samples. While South American sources show, as expected, a progressive increase in Fe oxidation with decreasing latitude, Antarctic sources show Fe oxidation levels higher than expected in such a cold polar environment, probably because of their very high exposure ages. Results from the TALDICE dust samples are compatible with a South American influence at the site during MIS2 (marine isotopic stage 2, the last and coldest phase of the last glacial period), in particular from Patagonia and Tierra del Fuego. However, a contribution from Australia and/or local Antarctic sources cannot be ruled out. Finally, important changes also occurred during the deglaciation and in the Holocene, when the influence of Antarctic local sources seems to have become progressively more important in recent times. This research is the first successful attempt to extract temporal climatic information from X-ray absorption spectroscopic data of the insoluble mineral dust particles contained in an ice core and shows the high potential of this technique.

Keywords: mineral dust; XANES; paleoclimatology; ice cores; southern hemisphere

---

1. Introduction

Mineral dust presents important climatic effects, both direct and indirect in relation to the atmospheric radiative budget [1–4] and with important influences on marine biogeochemical cycles and on the global carbon cycle [5]. During the last climatic cycles, the concentration of dust in the Antarctic ice changed up to 50 times between interglacial-to-glacial periods [6,7] an amount that corresponds to ~25 times in term of depositional flux [8]. The dust concentration dropped to extremely low levels (ppb level) during the Holocene and earlier interglacial thus making dust liquid counting and geochemical analyses extremely challenging [9]. Previous studies based on Sr and Nd radiogenic isotopes on several East Antarctic ice cores from the inner part of the polar plateau concluded that southern South America—in particular Patagonia, including Tierra del Fuego, as well as the southern part of central western Argentina [10–12], was the major dust supplier for central East Antarctica during Marine Isotope Stage (MIS) 2 [8,13–15] and likely earlier glaciations. These conclusions have been supported by Pb isotope data [13], and major and rare earth elements [10] analyses on the EPICA Dome C ice core. According to these studies, dust source regions for the central part of the EAIS are definitely homogeneous and well-defined during cold glacial periods. Open questions remain for interglacials, when the geochemical signatures are much more difficult to interpret, and a mixture of different remote sources is likely to have occurred [12].

Talos Dome is an area located at the margin of the East Antarctic Plateau (see Figure 1), close to the Ross Ice Sea and to the Southern Pacific Ocean, where the TALDICE (TALos Dome Ice CorE drilling project) ice core was recovered (72°49′ S–159°11′ E, 2,315 m a.s.l). The limited distance from the Southern Ocean and Ross Sea is responsible for the relatively high snow accumulation rate. The modern values of snow accumulation, in the order of 80 mm water equivalent per year (average 2004–1,259 AD, [16]) are relatively high compared to the inner Antarctica and for this reason highly resolved climatic records have been obtained from this core [11]. A peculiar geographical feature of Talos Dome in relation to the atmospheric dust cycle, is the presence of several sizeable ice-free sites in the Northern Victoria Land at only a few hundred km from Talos Dome (exposed moraines, raised beaches, regoliths and glacial deposits) some of which are located at high elevation and could protrude from the ice surface, remaining continuously ice-free since million years [17]. The presence of a coarse dust fraction (particles larger than 10 µm) in the atmospheric dust at Talos Dome [17] that is lacking in the interior of Antarctica, is a strong evidence that local sources play an important role in this marginal region of the ice sheet. The properties of dust in ice cores provide crucial information about the geochemistry and geology of the dust sources, in addition they can also reveal details about their environmental conditions during different climatic periods. However, the very low concentration of mineral particles in Antarctic ice, and their small size represent limitations to the application of well-established geochemical tools typically used in provenance studies.

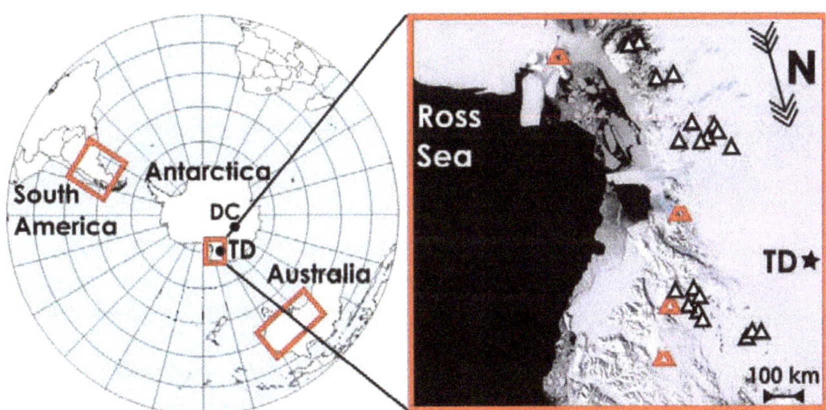

Figure 1. Map of the possible source areas in South America, Australia and Antarctica, the geographical areas considered in this study. TD stands for *Talos Dome* and DC for *Dome C*. On the right a zoom of the Antarctic region surrounding TD with highlighted the positions of the collection sites of the Antarctic potential source areas (PSA) (triangles) and the principal volcanoes (red cones).

Geochemical techniques such as Thermo-Ionization Mass Spectrometry (TIMS), Proton-Induced X-Ray and Gamma-ray Emission (PIXE-PIGE) techniques and low-background Instrumental Neutron Activation (INAA) were successfully applied to bulk dust from Antarctic ice cores [15,18,19]. Also, single-grain analyses can be applied to ice core samples. Electronic microscope with energy dispersive probe (SEM-EDAX) and Raman spectroscopy provide information about minerals and polymorphs, that are important to infer clues about the environmental conditions under which the sediment transported to the considered site, was produced [20]. Among the major elements, iron is particular important when considering the relationships existing between the biogeochemical cycles, dust, and climate. Indeed, iron associated with mineral dust plays a significant role in controlling the Southern Ocean primary bio-productivity and thus it directly influences the global carbon cycle, with relevant climatic consequences [21–23]. Iron concentration in Antarctic ice is normally at the level of $10^{-12} \sim 10^{-9}$ $g_{el} \cdot g^{-1}_{ice}$ (grams of element/grams of ice) [24,25], and for this reason its detection is a challenging analytical issue. However, not only concentration matters, there is another essential factor that must be considered when attention is given to the biogeochemical significance of iron: its speciation. The latter, together with the total amount of iron deposited at the ground in association to dust, influences the oceanic bio-productivity. Indeed, the Fe speciation is directly related to its solubility and thus to its bioavailability [26,27]. A major factor capable of influencing such geochemical features is mineralogy, but at present few studies have focused on the construction of ice core records about iron mineralogy and speciation, mostly because of the challenging analytical difficulties.

One of the most important features that influences iron bioavailability is its oxidation state. In minerals and magmas, iron typically occurs as Fe^{2+} and Fe^{3+}. Considering the superficial crustal environment, the two species are present in different contexts. Fe^{2+} is indicative of relatively fresh and unaltered rock outcrops, not subjected to intense chemical weathering. The reason is that Fe^{2+} is not stable in presence of atmospheric oxygen (it is oxidized to Fe^{3+}), therefore rocks exposed for a prolonged time to the atmosphere are generally depleted in Fe^{2+} and enriched in Fe^{3+}. The latter is dominant where chemical weathering is important and involves the exposition to the atmosphere. For example, in soils or outcrops with a prolonged exposure history [28]. The degree of oxidation of mineral Fe in dust samples could be thus potentially used to evaluate the degree of alteration and iron speciation of dust samples from potential source areas (PSA) and from an Antarctic ice core, to infer novel clues on the dust transport and deposition history in Antarctic during different climatic stages. The primary aim of this work is to demonstrate the feasibility of such an approach. We present here the first characterization concerning the Fe oxidation conditions of insoluble mineral dust particles extracted from ice samples of the Antarctic TALDICE ice core and from PSA samples of the southern hemisphere (Figure 1). After setting up the method [27,28] the goal has been reached through the analysis of the Fe K-edge XANES (X-ray absorption near edge structure) spectra of samples prepared following the procedure described in Section 4.

2. Materials and Methods

2.1. Ice Core Dust Samples

Given the extremely low concentration of dust in Antarctic ice, contamination issues are of primary importance when dealing with such samples. To limit as much as possible the contamination during preparation and measurement, it was necessary to develop a dedicated analytical protocol. Ice cores were cut in a cold room (T < −20 °C) at the Eurocold Laboratory (DISAT, University Milano-Bicocca) and decontaminated with repeated baths in ultra-pure water in an ISO6 class clean room, under an ISO5 laminar flow bench [6]. After melting mineral dust was extracted from meltwater through filtration, using pre-cleaned-acid rinsed [29]) polycarbonate membranes (pore size 0.4 µm). For each membrane, an appropriate amount of meltwater was filtered, to obtain 2–10 µg of dust. After filtration, the membranes were mounted on specifically designed PTFE sample holders and sealed in clean plastic containers until the measurement at the synchrotron facility. For this study, a total of 44 samples were prepared, covering the time period between 2 and 25 kyrs BP, corresponding to the last glacial maximum (LGM, 25–18 kyrs BP), the deglaciation (or termination I, 18–11.7 kyrs BP) and the current interglacial period, the Holocene (11.7 kyrs BP—present).

2.2. PSA Samples

Despite it is relatively recent, elemental speciation with the XANES method is well established in the field of geo- and environmental chemistry [30]. Through this technique it is possible to investigate many elements, but Fe plays a relevant role and many XANES-based studies focused on it [31–35]. Among the geochemical features that is possible to retrieve by XANES, iron oxidation state can be inferred from different spectral features related to the K-edge absorption properties: the pre-edge peak [36] or even from the shift of the energy of the main absorption jump [37]. Since the analysis of the pre-edge strongly depends on the resolution and on the S/N ratio, in this work we focused on the analysis of the edge energy. Its position depends on two main factors: the oxidation state of Fe (i.e., the Fe^{3+}/Fe^{2+} ratio) and its coordination [37]. However, as discussed in the next paragraphs, dealing with atmospheric dust and PSA, the iron coordination in the inorganic component can be considered only octahedral, thus the position of the K adsorption edge is mostly related to changes in the relative concentration of iron Fe^{3+} and Fe^{2+}. Indeed, considering upper crustal samples, all the most common iron oxides display a coordination number of six, i.e., it can be assumed that Fe is almost present in its octahedral coordination state [27,35,36].

A total of 74 PSA samples were considered in this work. They were collected from Australia (17), Antarctica (Victoria Land, 20) and South America (34) plus three samples of NIST standard soil. The choice of considering these areas, stems from previous results about the provenance of dust transported to Antarctica in different climatic periods. Several studies suggested South America as the dominant source for the whole East Antarctic Ice Sheet (EAIS) during glacial periods, including the LGM considered in this work [6,8,14]. Australia was suggested as a significant source for the dust deposited in inner EAIS during interglacials [10,19]. Finally, local Antarctic dust sources located in the Victoria Land region were considered since several evidence pointed to a regionalization of the dust cycle in the Ross Sea area during the Holocene [19,21]. For the details about the collection sites and their morphological and environmental features the reader is referred to previous works where they are fully presented and discussed [7,10,17]. Samples considered here are the same considered in [21], where attention was given to other geochemical features (Sr-Nd isotopic composition). The only samples that are here considered for the first time are those from Tierra del Fuego, collected from deflation surfaces and aeolian deposits, from 53° S to 49° S between Chile and Argentina on both sides of the Magellan Strait. PSA samples were prepared to be as similar as possible to ice core dust ones. A gravimetric wet method was used to extract the size fraction below 5 μm, i.e., the one subjected to long range atmospheric transport [6].

2.3. XANES Analysis

In the 70's the absorption peaks near the X-Ray absorption edges have been shown to arise from shape resonances of the excited photoelectron confined by multiple scattering within a nanoscale cluster centered at the absorbing atomic species in disordered oxides [38] and called with the acronym XANES [39]. Nowadays this spectroscopy is widely used to investigate the site specific electronic and the local geometrical structure (i.e., coordination and bond angles) of nanoscale clusters surrounding the selected absorber atom thanks to the multiple scattering (MS) data analysis in the real space [40]. This technique provides a unique tool to identify the different contributions in the spectra of heterogeneous systems containing several iron oxides [41,42].

XANES experiments were performed at the Diamond Lightsource, on beamline B18, dedicated to X-ray absorption spectroscopy [43]. It was necessary to adopt several expedients to limit sample contamination during the experimental runs. At first a sealed plastic glove box was mounted in direct contact to the experimental chamber of the beamline. Secondarily, clean plastic sheets were applied to the inner walls of the experimental chamber, to limit the inelastic scattering given by the interaction between the incident photons and the metallic walls and reduce the associated background. To further reduce the background signal, measurements were carried out at high-vacuum conditions to exclude the contribution from atmospheric gases. The acquisition of the spectra was done using a Vortex

4-elements silicon-detector. Given its large collection surface, the high resolution (140 eV FWHM at the 5.9 keV Mn Kα line), and the sensitivity for low energy fluorescence and high peak-to-background ratio, it proved to be suited for our low concentration samples. For each sample, the absorption spectrum related to the Fe K-edge transition was acquired three times and averaged. Each acquisition lasted for 0.5–1 h, depending on the sample concentration. The energy interval between 6900 and 7800 eV was considered and the average resolution of the spectra is 0.2 eV (Figure 2). Further details can be found in previous works [44,45].

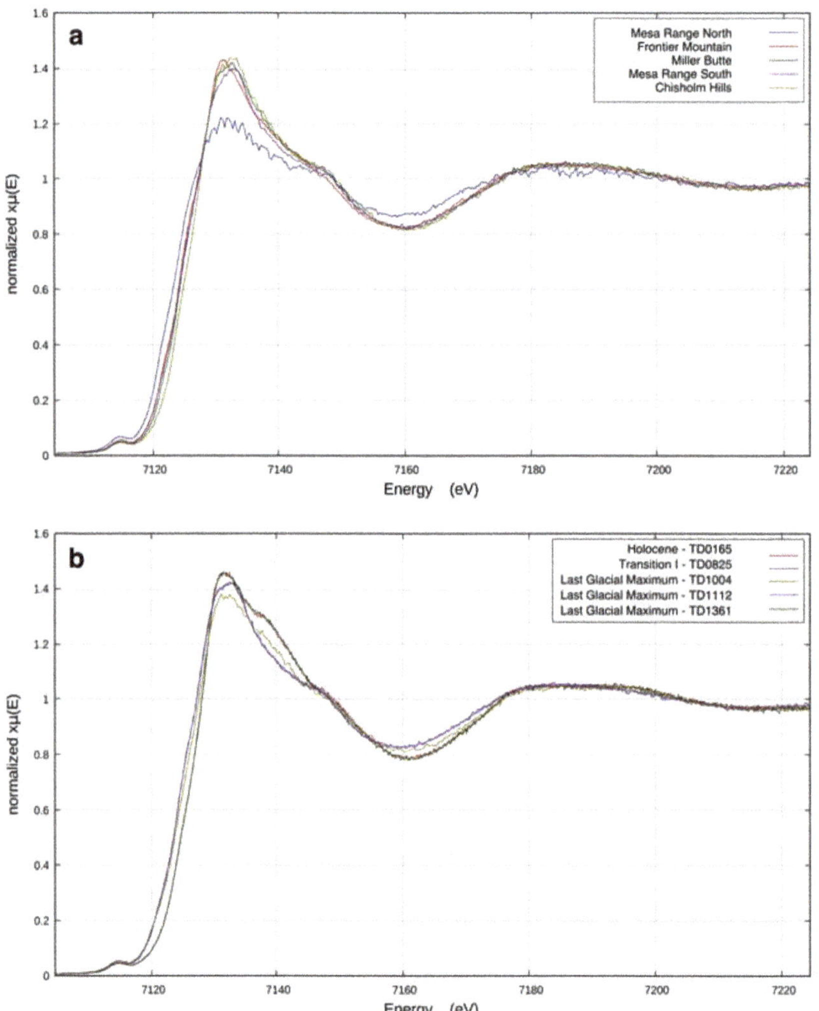

Figure 2. (**a**) Comparison of Fe K-edge XANES spectra of soils from different local Antarctic sources; (**b**) and Fe K-edge spectra from different periods along the TD ice core.

Spectra were normalized using the pre-edge baseline, and the Fe K-edge (E_0) was determined to be the inflection point, which is the maximum of the first derivative (see Figure 3 and Figure SM2) or the zero of the second derivative of the XANES spectrum. To compare spectra, they were normalized at the same energy. For the interpretation and analysis of the data we considered the energy corresponding to the normalized signal intensity of 0.8. This choice was made to consider a robust value, minimizing possible electronic contributions at the edge [37]. The accuracy of the method was monitored measuring between one sample and another, the XANES spectrum related to a metallic Fe foil. In this way it was possible to maintain the same energy calibration during the entire run. Conversely, precision was evaluated acquiring the Fe K-edge XANES spectrum of the NIST soil standard reference material (SRM 2709a, San Joaquin soil). It can be seen in Table 1 that the standard deviation of three successive acquisitions is 0.18 eV, a value lower than the detector spectral resolution.

Table 1. The main statistics parameters of PSA from the analysis of the Fe K-edge XANES spectra.

	Antarctica	Tierra Del Fuego	Patagonia	Pampa	Australia	NIST
N	18	7	14	12	17	3
Mean (eV)	7125.58	7124.74	7125.61	7126.57	7126.70	7126.25
Median (eV)	7125.53	7125.10	7125.70	7126.71	7126.73	7126.32
Std. Error	0.12	0.35	0.23	0.08	0.03	0.10
Variance	0.24	0.86	0.76	0.08	0.02	0.03
Std. Dev.	0.49	0.93	0.87	0.28	0.14	0.18
Coeff. Var.	0.006	0.013	0.012	0.004	0.002	0.003

Additional representative XANES spectra of PSA are showed in Figure 3 to demonstrate the variability of these spectra as determined by the mixture of different iron oxides present in the soils from different PSA. Some TD ice core samples are shown, together with their first derivative in Figures SM1 and SM2 in the Supplementary Materials.

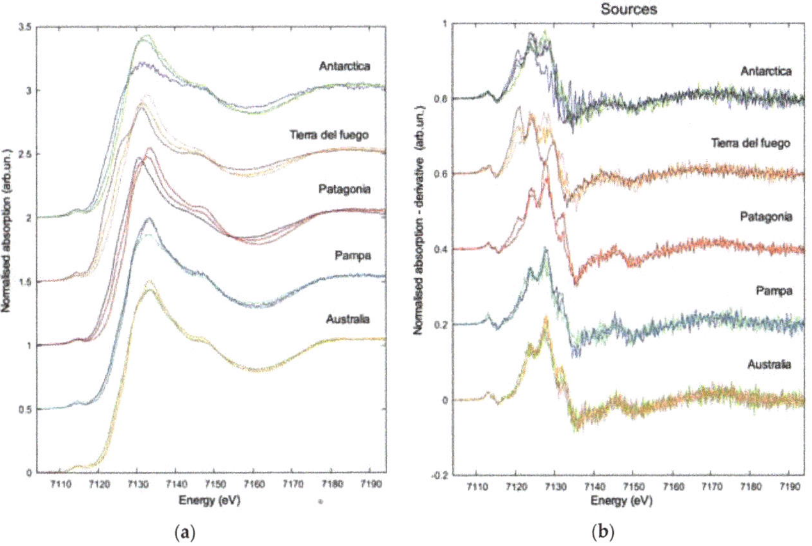

Figure 3. (**a**) Comparison of Fe K-edge XANES spectra for some representative spectra PSAs discussed in the text; (**b**) comparison of the first derivative of the XANES spectra at the Fe K-edge for the representative spectra of sources discussed in the text.

3. Results

3.1. The PSA Fe K-Edge Energy Measurements

A summary concerning the data about the Fe K-edge of the PSA samples is reported in Table 1, whereas the Student *T*-tests applied to data to highlight significant differences between the samples (see Table SM1 of the Supplementary Materials). Australia presents the higher Fe K-edge energy, reflecting a strong oxidation of Fe, an expected scenario due to the climatic and environmental conditions of this continent. Australian PSAs are related to temperate and tropical climatic conditions and highly stable surfaces, characterized by long exposure histories. Such features determine a strong chemical weathering and the occurrence of highly evolved outcrops and soils. Under similar conditions oxidation processes and the in situ formation of iron oxides are both favored [46,47].

Typical soils found in Southern Australia are oxisols, Alfisols podzols, prairie soils and red earth soils [http://www.clw.csiro.au/aclep/soilandlandscapegrid/]; their rusty colors directly reflect the presence of Fe^{3+}, in agreement with XANES spectra that show a high Fe K-edge energy (average value 7126.7 eV), compatible with an almost complete oxidation of the Fe fraction present in these soil samples. In addition to the high oxidation, another feature characterizing the Australian samples is the homogeneity. In these samples the Fe K-edge energy ranges from 7126.39 eV to 7126.93 eV, with a very low standard deviation: 0.14 eV (see also Figure 4).

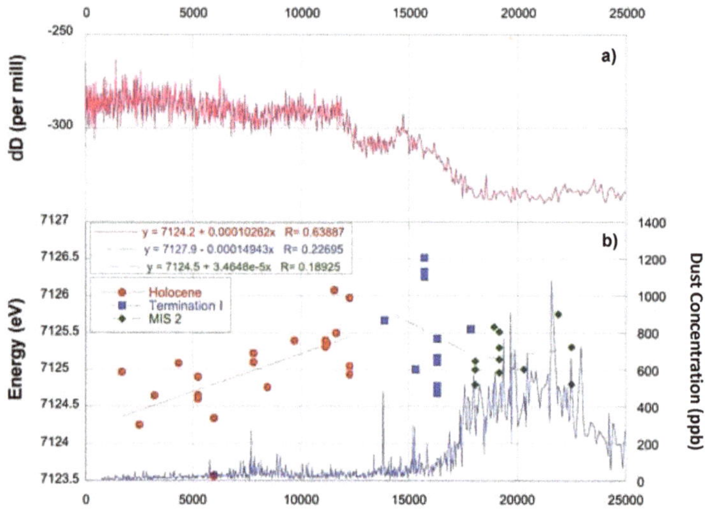

Figure 4. (**a**) the TD stable isotope record indicating the climatic changes between MIS2 and Holocene (pink line, [48]). (**b**) the TD dust mass concentration record (blue line [22,34]) and the Fe K-edge energy positions obtained through from the XANES analysis (circles, square, diamonds for Holocene, Termination I and MIS2 samples, respectively). In the insets, linear regressions are added, together with the equations and the correlation coefficients.

This is a relevant result since Australian PSA samples were collected in different geographic and climatic areas. On the contrary, South America PSA samples [49] are characterized by a wider range of Fe K-edge energy (from 7123.5 eV to 7127.0 eV). This evidence suggests a wide latitudinal extension of the PSA considered in this work (ranging from 30° S to 55° S), implying very different climatic regimes. For this reason, South America samples have been divided into three sub-groups related to their geographical provenance: Pampas—Central Argentina (30°–45° S), Patagonia (45°–52° S) and Tierra del Fuego (south of 52° S). Figure 4 clearly shows an inverse relationship between Fe oxidation and latitude among the South America samples, while those from Pampas-central Argentina regions (30° S to 45° S) display a Fe K-edge energy range from 7126 to 7127 eV. Patagonian samples (45° S to 52° S) falls in a wider energy range from 7124.8 eV to 7127 eV, while those of Tierra del Fuego (south of 52° S) show lower Fe K-edge energies, between 7123.5 eV and 7125.8 eV, suggesting conditions less favorable for oxidative processes. Such results show that the position of the Fe K-edge in South American PSA samples is strongly controlled by latitude and thus by the climatic conditions characterizing the different areas where samples were collected. The relatively high oxidation of the PSA samples from the Pampas—Central Argentina samples, comparable to the Australian ones, reflects the climatic and pedogenetic processes occurring in this region. The Pampa area is originating from soils produced by the strong alteration of acid parent materials (Andean rocks and volcanic products) and Mollisols and Alfisols. The latter are from the humid Pampa area where the pedogenetic processes are responsible for the processing formation of aluminum and iron oxides. Considering the arid mid-latitude area of the Central Argentinean plateau, Entisols and Aridisols are common, pointing to a slower, but active, oxides formation [50]. Despite the presence of soils that are very similar to the ones found on the central Argentina plateau, Patagonian samples present lower energies for the Fe K-edge, in accordance to less oxidized conditions. This could be related to the temperate-to-cold climatic conditions found in Patagonia. Here, the degree of chemical weathering is weakened by the low temperatures and reduces the oxidation processes [50]. Finally, the semi-arctic climate of Tierra del Fuego easily explains the occurrence of very low Fe K-edge energies that confirms poor chemical weathering and a larger amount of Fe^{2+}. It should be also considered that during the LGM, this region was extensively covered by a continental ice cap whose presence was responsible for the massive production and deposition of glacial and volcanic sediments. Given the cold climate and the recent deposition of such sediments, it is reasonable to associate these samples to a less weathered geochemical signature.

Focusing on the Antarctic PSA samples, it would have been expected to find very low values for the energy of the Fe K-edge, owing to the cold and dry climate of Victoria Land, not favorable to chemical weathering and to the development of soils conditions [17]. Cryoturbation and permafrost processes represent the only active mechanism [51]. XANES results partly support these considerations. Despite the average Fe K-edge energy of Antarctic PSA is actually the lower one among the considered regions (7125.5 eV, see also Figure 4 and Table 1), its variability is high, and some samples shows a moderately high energy, reflecting a relevant Fe oxidation. Only South America PSA showed a larger variability, but in this case a wide geographical area was considered, at variance of the Antarctic samples. All the Antarctic PSA were indeed collected in Victoria Land, representing the major area with free-ice of this Antarctic sector, within a few hundred km. To explain the high variability, it is necessary to consider additional factors. At first Antarctic PSA considered in this work are both 'primary' dust sources, for example regoliths, i.e., the primary product deriving from the alteration (chemical and mostly physical) of the many ice-free rocky outcrops characterizing Victoria Land, and 'secondary' sources. The latter consist in PSA samples obtained from reworked and mixed deposits, composed by sediments that were already subjected to transport, for example glacial drifts, and aeolian deposits, as it was extensively described in earlier studies [17]. In addition, we must underline that Victoria Land presents a complex geological history, resulting in the outcropping of many different lithologies within a few km. A final, but important feature, is the long exposure history of many ice-free sites of Victoria Land, that in some cases remained deglaciated for several million years. Such long exposure ages and the occasional occurrence of favorable conditions for chemical weathering [17,52] could explain the high Fe K-edge energy displayed by some of these Antarctic samples.

3.2. Talos Dome Mineral Dust

The dust record of the Talos Dome ice core over the last 25 ky reflect the effects related to the climatic transition from the last glacial period to the current interglacial one, the Holocene. The effects on the dust cycle are well known: a drastic decrease of the atmospheric dust burden. It was related to environmental and atmospheric changes that affected the southern hemisphere during this period [20]. At Talos Dome the dust depositional flux during the LGM was 6–7 times higher than in the Holocene [19]. As for the EPICA Dome C ice core, typical Holocene dust concentrations in ice at Talos Dome were reached around 14.6 ky BP, just before the Antarctic Cold Reversal, about 3000 years before the effective onset of the Holocene, likely because of changes in the hydrologic cycle [6]. Interestingly, the position of the Fe K-edge energy of Talos Dome ice core dust samples does not follow the dust concentration during the last 25 ky (Figure 4a), but is more likely correlated with the assemblage of Fe-bearing minerals that determining the final shape of the XANES spectra also tune the energy at the Fe K-edge. Data are summarized in Table 2, whereas the Student T-tests are reported in the Table SM2 in the Supplementary Materials.

MIS2 presents a low variability of the Fe K-edge (mean energy 7125.19 ± 0.09 eV), despite dust mass concentration changes measured in this period (between 300 and 800 ng g^{-1}). This evidence points to a homogeneous dust composition in term of Fe content minerals and supports the hypothesis that during glacial periods dust transported to Antarctica is uniform at the continental scale [20].

Table 2. The statistical parameters related to the energy position of the Fe K-edge in the Talos Dome ice core mineral dust samples for the different climatic periods considered in this work.

	Holocene	Termination I	MIS2
Samples	21	11	12
Mean (eV)	7125.00	7125.50	7125.19
Median (eV)	7125.05	7125.42	7125.13
Std. Deviation	0.57	0.63	0.30
Variance	0.32	0.40	0.09
Std. Error	0.12	0.19	0.09
Correlation Coeff. R	0.64	0.23	0.19

The end of MIS2 occurred at around 18 ky BP, corresponding to the onset of the last deglaciation, also known as Termination I (Figure 4a). This is an intermediate climatic period when changes from full glacial to interglacial conditions occurred. The relevant climatic and environmental changes that characterized this period are also confirmed by the dust concentration record of the Talos Dome ice core, which shows a significant decrease of the concentration (Figure 4b). Also, the Fe K-edge energy was affected by the climatic transition. In this period, it shows a high variability and slightly higher average values with respect to MIS2 (mean: 7125.50 ± 0.63 eV).

During the Holocene, i.e., the current interglacial, the transition edge decreases quite uniformly during the entire climatic period. From early to late Holocene a linear decreasing trend is recognized (Figure 4b). The variance (0.32) and standard error (0.12) of the iron K-edge energy position remain high also during the Holocene (Table 2). It must be noted that this could be related to the extremely low dust concentration during this period. However, it is evident that a change from more to less oxidized iron-bearing mineral phases occurred during the Holocene and the last part of Termination I (the last 13,500 years).

4. Discussion

The energy of the Fe K-edge of PSA samples shows a clear difference between the less weathered areas of Antarctica and the more oxidized Australian samples, reflecting the strong climatic differences of these southern hemisphere regions (Figure 5b). Not only PSA, but also Talos Dome mineral dust samples present a relevant variability with respect to the Fe mineral composition, in particular when considering the three climatic periods that define the last 25 kyrs (Figure 5a and Student *T*-tests in Table SM3 in the Supplementary Materials). In the last glacial period the Fe oxidation of TALDICE dust is compatible with a South American provenance from Patagonia and Tierra del Fuego (the average iron K-edge energy for the Talos Dome MIS2 dust is 7125.2 eV, while the average for the PSA America origins from Patagonia and Tierra del Fuego is 7124.7 and 7125.6 eV respectively) although we cannot rule out also the Antarctic sources, owing to a mean K-edge transition that is very close to the one of southern South America samples (average value 7125.5 eV). When complementary data are considered, it is very likely that the Antarctic local contribution was actually present during MIS2, also considering that many Antarctic PSAs were ice-free also during this period [17,52], being probably negligible considering the intense contribution from South America. During Termination I the transport of dust from South America to Antarctica decreased, mainly because of changes in the hydrological cycle. The iron K-edge energy in this period shows evident changes. The beginning of Termination I is characterized by an increase of Fe oxidation (increase of the Fe K-edge energy), decreasing in the next period, and this trend is maintained during the entire Holocene.

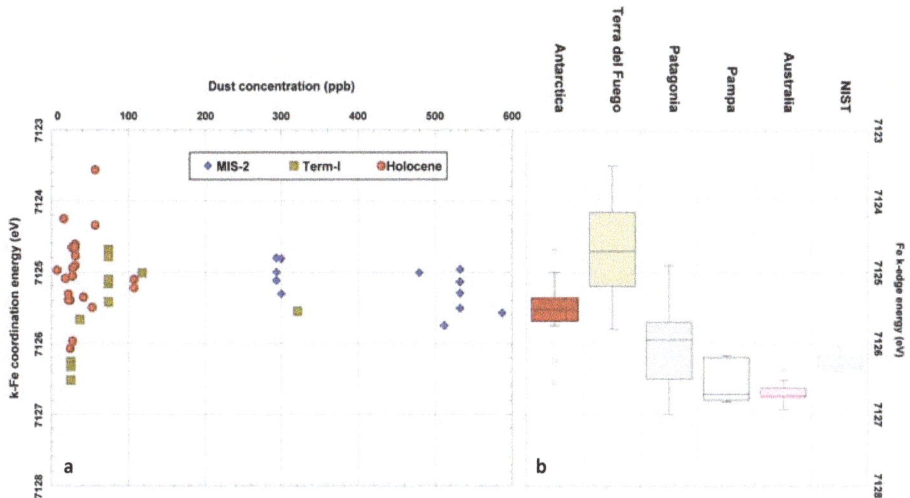

Figure 5. (a) The mineral dust concentration vs. Fe K-edge energy for different climatic periods of TD samples, and (b) the box-plot of the Fe K-edge energy of the different PSA samples compared to the NIST standard.

The position of the Fe K-edge shift obtained considering the linear fit shown in Figure 4b changes from 7125.5 eV at the beginning of the Holocene (12 Kyr BP), to less oxidized values during late Holocene (7124.4 eV at 2.5 Kyr BP) when the local dust contribution became more relevant. It suggests that the shift to interglacial conditions determined a first increase of the Fe oxidation, probably related to a first mobilization of the glacial deposits accumulated during MIS2 and heavily reworked. Successively the position of the edge starts decreasing, and the trend persists across the entire Holocene. It is responsible for the decrease of the Fe K-edge from the maximum value of 7126.1 eV at 12 ky BP to the minimum value of 7124.2 at 2.5 ky BP. The lower value (7123.55 eV) is from one sample at 6 ky BP, with a possible volcanic contamination. Two hypotheses can interpret these findings:

1. the geochemical properties of the dust deposited at TD change in relation to environmental and climatic modifications of the PSA [44];
2. the shift from glacial to interglacial conditions impacted the dust cycle and the atmospheric circulation at TD, with changes in terms of dust provenance and of the relative contribute from different PSA.

The climatic shift from cold and dry conditions of MIS2 to the interglacial period, characterized by a wetter and warmer climate, affected the production and emission of dust over the continents, with further consequences concerning the pedogenetic processes. Interglacial conditions, thanks to the reinforcement of the hydrological cycle and to an increase of temperatures, favor chemical alterations and oxidative processes with an expected increase of the Fe K-edge energy. If the first hypothesis was correct, we would have appreciated an increase of the Fe K-edge as a consequence of the last climatic transition, with a strong increasing oxidation (higher transition energies), characterizing the Talos Dome Holocene dust. This is exactly the opposite of what has been actually observed (Figure 4b). Consequently, the second hypothesis based on the change in the relative weight of the different sources, appears the best scenario to explain the Holocene presented in Figure 4b. In agreement with earlier studies [19,21] we suggest that during the deglaciation and Early Holocene a progressive change in terms of active dust sources with respect to TD occurred. While the transport of dust from South America progressively decreased, local Antarctic sources gained importance and became dominant, with possible but minor contributions also from Australia, in agreement with climate models [45]. This second scenario can thus explain the less oxidized Fe signature observed during the Holocene with lower Fe K-edge energies, as expected for dust emitted by high altitude glacial deposits and regoliths from Northern Victoria Land, such as those present in the Mesas outcropping area [21]. The role played by local Antarctic sources is also supported by the dust grain size distributions observed in Talos Dome ice core samples (Figures 4 and 5). They are containing local coarse particles (5–10 μm diameter). Although present both during MIS2 and Holocene, these coarse particles became more abundant in relative terms during the Holocene and especially during late Holocene, when they account for 50–70% of total dust mass [17].

The possible role played by in situ oxidation of Fe minerals within ice is not discussed here, despite preliminary results showed its relevance only in very deep ice, in agreement with what already observed in the EPICA Dome C ice core [53] and, in the case of TALDICE below 1400 m [54].

In summary, the XANES-based Fe speciation analysis permits to evaluate the different contributions of mineral dust from the possible source areas that supply dust to the East Antarctic plateau. Some difficulties remain to distinguish unambiguously the contribution of some geographically different sources with similar Fe oxidation conditions, and only a comprehensive mineralogical study may help to clarify the role of different sources. However, as already demonstrated by many researches on mineral samples containing iron atoms in both tetrahedral and octahedral sites a clear energy shifts by 2–3 eV to high energy is observed increasing the Fe^{3+} occupancy. Moreover, as shows in Ref. [48] the iron T-O distances are typically much lower than 2 Å (1.65–1.66 Å) while octahedral distances are greater than 2 Å (2.06–2.11 Å). In addition, to the different potentials due to the symmetry, a tetrahedral iron coordination should give a strong pre-edge feature while the difference in distance of 20–25% should also imply a clear shift of the XANES features. Consequently, in our samples the fractional presence of tetrahedral coordinated iron would give rise to a significant shift in the edge position and of the XANES features. The joint analysis of pre-edge and XANES features indicates that tetrahedral/non-centro-symmetrical components, if present, are small or negligible. A discussion on the analysis of the pre-edge of iron in octahedral site in minerals has been recently published [55].

Despite the limitations, spectroscopic data presented in this contribution support earlier evidence of a South American provenance for the dust deposited in peripheral East Antarctica during MIS2 and highlights a progressive enhancement in the relative proportion of local dust sources during the Holocene. Although preliminary, this research is the first attempt to extract climatic information from XANES spectroscopic data obtained from the analysis of insoluble mineral dust particles contained in an ice core. Although other methods such as the analysis of the pre-edge: centroid and/or fit, represent powerful approach that can be used for the Fe speciation, at present we cannot obtain a reliable Fe K-edge dataset to compare with the high concentrated PSA samples because of the very low concentration and the complex composition of the mineral dust trapped in the Antarctic ice. For the same reason, the possibility to extract the Fe_{tot}/Fe^{3+} ratio is not yet possible at this stage.

Supplementary Materials: The following are available online at http://www.mdpi.com/2410-3896/3/4/45/s1.

Author Contributions: V.M., G.B., B.D. performed the TALDICE paleoclimatic interpretation and sources/ice core relationships; C.M., G.C., D.H. performed the experiments at Diamond, and data analysis of XAFS spectra. All authors wrote the manuscript and contributed to the discussion/interpretation of the results.

Funding: Part of this research has been also founded by Italian National Antarctic Program project n° PNRA 2013/B2.10, PI S. Rocchi (UNIPI). The support of DARA Department of the Italian Presidenza del Consiglio dei Ministri is gratefully acknowledged.

Acknowledgments: This research is a collaboration made among Milano-Bicocca University, INFN-Laboratori Nazionali di Frascati, and the Diamond Light Source facility. Part of the sample preparation was performed at the EuroCold Lab at University of Milano-Bicocca, founded by Italian National Science Foundation Next Data Project. XANES spectra were collected in the framework of Proposals 90U5 and 3082M at SSRL, a national user facility operated by Stanford University on behalf of the U.S. Department of Energy, Office of Basic E he iron T-O distances are typically mnergy Sciences. XANES measurements were also performed at Diamond, the UK national synchrotron radiation facility in the framework of the proposal NT1984 and at ESRF within the experiment 08-01-1031 on Beamline BM08. We strongly acknowledge the staff of the Italian CRG LISA for support of XANES measurements at the beamline BM08 at ESRF within the experiment 08-01-1031 and for many fruitful discussions.

Conflicts of Interest: The authors declare no conflict of interest. The founding sponsors had no role in the design of the study; in the collection, analyses, or interpretation of data; in the writing of the manuscript, and in the decision to publish the results.

References

1. Maher, B.A.; Prospero, J.M.; Mackie, D.; Gaiero, D.; Hesse, P.P.; Balkanski, Y. Global connections between aeolian dust, climate and ocean biogeochemistry at the present day and at the last glacial maximum. *Earth-Sci. Rev.* **2010**, *99*, 61–97. [CrossRef]
2. Maggi, V. Mineralogy of atmospheric microparticles deposited along the Greenland ice core project ice core. *J. Geophys. Res. Oceans* **1997**, *102*, 26725–26734. [CrossRef]
3. Maggi, V.; Petit, J.R. Atmospheric dust concentration record from the Hercules neve firn core, northern Victoria land, Antarctica. *Ann. Glaciol.* **1998**, *27*, 355–359. [CrossRef]
4. Maggi, V.; Villa, S.; Finizio, A.; Delmonte, B.; Casati, P.; Marino, F. Variability of anthropogenic and natural compounds in high altitude-high accumulation alpine glaciers. *Hydrobiologia* **2006**, *562*, 43–56. [CrossRef]
5. Jickells, T.; Boyd, P.; Hunter, K.A. Biogeochemical impacts of dust on the global carbon cycle. In *Mineral Dust*; Knippertz, B., Stuut, J.B., Eds.; Springer: Dordrecht, The Netherlands, 2014; pp. 359–384.
6. Delmonte, B.; Basile-Doelsch, I.; Petit, J.; Maggi, V.; Revel-Rolland, M.; Michard, A.; Jagoutz, E.; Grousset, F. Comparing the epica and vostok dust records during the last 220,000 years: Stratigraphical correlation and provenance in glacial periods. *Earth-Sci. Rev.* **2004**, *66*, 63–87. [CrossRef]
7. Delmonte, B.; Andersson, P.; Hansson, M.; Schoberg, H.; Petit, J.; Basile-Doelsch, I.; Maggi, V. Aeolian dust in east Antarctica (epica-dome c and vostok): Provenance during glacial ages over the last 800 kyr. *Geophys. Res. Lett.* **2008**, *35*. [CrossRef]
8. Lambert, F.; Delmonte, B.; Petit, J.R.; Bigler, M.; Kaufmann, P.R.; Hutterli, M.A.; Stocker, T.F.; Ruth, U.; Steffensen, J.P.; Maggi, V. Dust—Climate couplings over the past 800,000 years from the Epica dome c ice core. *Nature* **2008**, *452*, 616–619. [CrossRef]
9. Delmonte, B.; Petit, J.R.; Krinner, G.; Maggi, V.; Jouzel, J.; Udisti, R. Ice core evidence for secular variability and 200-year dipolar oscillations in atmospheric circulation over east Antarctica during the Holocene. *Clim. Dyn.* **2005**, *24*, 641–654. [CrossRef]
10. Revel-Rolland, M.; De Deckker, P.; Delmonte, B.; Hesse, P.; Magee, J.; Basile-Doelsch, I.; Grousset, F.; Bosch, D. Eastern Australia: A possible source of dust in east Antarctica interglacial ice. *Earth Planet. Sci. Lett.* **2006**, *249*, 1–13. [CrossRef]
11. Frezzotti, M.; Pourchet, M.; Flora, O.; Gandolfi, S.; Gay, M.; Urbini, S.; Vincent, C.; Becagli, S.; Gragnani, R.; Proposito, M.; et al. Spatial and temporal variability of snow accumulation in east Antarctica from traverse data. *J. Glaciol.* **2005**, *51*, 113–124. [CrossRef]
12. Scarchilli, C.; Frezzotti, M.; Ruti, P. Snow precipitation at four ice core sites in east Antarctica: Provenance, seasonality and blocking factors. *Clim. Dyn.* **2011**, *37*, 2107–2125. [CrossRef]
13. Vallelonga, P.; Gabrielli, P.; Balliana, E.; Wegner, A.; Delmonte, B.; Turetta, C.; Burton, G.; Vanhaecke, F.; Rosman, K.; Hong, S.; et al. Lead isotopic compositions in the epica dome c ice core and southern hemisphere potential source areas. *Quat. Sci. Rev.* **2010**, *29*, 247–255. [CrossRef]
14. Basile, I.; Grousset, F.; Revel, M.; Petit, J.; Biscaye, P.; Barkov, N. Patagonian origin of glacial dust deposited in east Antarctica (Vostok and Dome c) during glacial stages 2, 4 and 6. *Earth Planet. Sci. Lett.* **1997**, *146*, 573–589. [CrossRef]
15. Grousset, F.; Biscaye, P.; Revel, M.; Petit, J.; Pye, K.; Joussaume, S.; Jouzel, J. Antarctic (dome c) ice-core dust at 18 ky bp—Isotopic constraints on origins. *Earth Planet. Sci. Lett.* **1992**, *111*, 175–182. [CrossRef]
16. Frezzotti, M.; Bitelli, G.; De Michelis, P.; Deponti, A.; Forieri, A.; Gandolfi, S.; Maggi, V.; Mancini, F.; Remy, F.; Tabacco, I.; et al. Geophysical survey at talos dome, east antarctica: The search for a new deep-drilling site. *Ann. Glaciol.* **2004**, *39*, 423–432. [CrossRef]

17. Delmonte, B.; Baroni, C.; Andersson, P.; Schoberg, H.; Hansson, M.; Aciego, S.; Petit, J.; Albani, S.; Mazzola, C.; Maggi, V.; et al. Aeolian dust in the talos dome ice core (east Antarctica, pacific/ross sea sector): Victoria land versus remote sources over the last two climate cycles. *J. Quat. Sci.* **2010**, *25*, 1327–1337. [CrossRef]

18. Baccolo, G.; Clemenza, M.; Delmonte, B.; Maffezzoli, N.; Nastasi, M.; Previtali, E.; Prata, M.; Salvini, A.; Maggi, V. A new method based on low background instrumental neutron activation analysis for major, trace and ultra-trace element determination in atmospheric mineral dust from polar ice cores. *Anal. Chim. Acta* **2016**, *922*, 11–18. [CrossRef]

19. Marino, F.; Calzolai, G.; Caporali, S.; Castellano, E.; Chiari, M.; Lucarelli, F.; Maggi, V.; Nava, S.; Sala, M.; Udisti, R. Pixe and pige techniques for the analysis of antarctic ice dust and continental sediments. *Nuclear Instrum. Methods Phys. Res. Sect. B-Beam Interact. Mater. Atoms* **2008**, *266*, 2396–2400. [CrossRef]

20. Delmonte, B.; Paleari, C.; Ando, S.; Garzanti, E.; Andersson, P.; Petit, J.; Crosta, X.; Narcisi, B.; Baroni, C.; Salvatore, M.; et al. Causes of dust size variability in central east Antarctica (dome b): Atmospheric transport from expanded south American sources during marine isotope stage 2. *Quat. Sci. Rev.* **2017**, *168*, 55–68. [CrossRef]

21. Jickells, T.; An, Z.; Andersen, K.; Baker, A.; Bergametti, G.; Brooks, N.; Cao, J.; Boyd, P.; Duce, R.; Hunter, K.; et al. Global iron connections between desert dust, ocean biogeochemistry, and climate. *Science* **2005**, *308*, 67–71. [CrossRef]

22. Mahowald, N.; Baker, A.; Bergametti, G.; Brooks, N.; Duce, R.; Jickells, T.; Kubilay, N.; Prospero, J.; Tegen, I. Atmospheric global dust cycle and iron inputs to the ocean. *Glob. Biogeochem. Cycles* **2005**, *19*. [CrossRef]

23. Blain, S.; Queguiner, B.; Armand, L.; Belviso, S.; Bombled, B.; Bopp, L.; Bowie, A.; Brunet, C.; Brussaard, C.; Carlotti, F.; et al. Effect of natural iron fertilization on carbon sequestration in the southern ocean. *Nature* **2007**, *446*, 1070–1071. [CrossRef] [PubMed]

24. Gaspari, V.; Barbante, C.; Cozzi, G.; Cescon, P.; Boutron, C.; Gabrielli, P.; Capodaglio, G.; Ferrari, C.; Petit, J.; Delmonte, B. Atmospheric iron fluxes over the last deglaciation: Climatic implications. *Geophys. Res. Lett.* **2006**, *33*. [CrossRef]

25. Edwards, R.; Sedwick, P.; Morgan, V.; Boutron, C. Iron in ice cores from law dome: A record of atmospheric iron deposition for maritime east antarctica during the holocene and last glacial maximum. *Geochem. Geophys. Geosyst.* **2006**, *7*. [CrossRef]

26. Spolaor, A.; Vallelonga, P.; Cozzi, G.; Gabrieli, J.; Varin, C.; Kehrwald, N.; Zennaro, P.; Boutron, C.; Barbante, C. Iron speciation in aerosol dust influences iron bioavailability over glacial-interglacial timescales. *Geophys. Res. Lett.* **2013**, *40*, 1618–1623. [CrossRef]

27. Schroth, A.; Crusius, J.; Sholkovitz, E.; Bostick, B. Iron solubility driven by speciation in dust sources to the ocean. *Nat. Geosci.* **2009**, *2*, 337–340. [CrossRef]

28. Wedepohl, K. The composition of the continental-crust. *Geochim. Cosmochim. Acta* **1995**, *59*, 1217–1232. [CrossRef]

29. Baccolo, G.; Maffezzoli, N.; Clemenza, M.; Delmonte, B.; Prata, M.; Salvini, A.; Maggi, V.; Previtali, E. Low-background neutron activation analysis: A powerful tool for atmospheric mineral dust analysis in ice cores. *J. Radioanal. Nuclear Chem.* **2015**, *306*, 589–597. [CrossRef]

30. Walker, S.; Jamieson, H.; Lanzirotti, A.; Andrade, C.; Hall, G. The speciation of arsenic in iron oxides in mine wastes from the giant gold mine, nwt: Application of synchrotron micro-xrd and micro-xanes at the grain scale. *Can. Mineral.* **2005**, *43*, 1205–1224. [CrossRef]

31. Bajt, S.; Sutton, S.; Delaney, J. X-ray microprobe analysis of iron oxidation-states in silicates and oxides using X-ray-absorption near-edge structure (xanes). *Geochim. Cosmochim. Acta* **1994**, *58*, 5209–5214. [CrossRef]

32. Galoisy, L.; Calas, G.; Arrio, M. High-resolution xanes spectra of iron in minerals and glasses: Structural information from the pre-edge region. *Chem. Geol.* **2001**, *174*, 307–319. [CrossRef]

33. Strawn, D.; Doner, H.; Zavarin, M.; McHugo, S. Microscale investigation into the geochemistry of arsenic, selenium, and iron in soil developed in pyritic shale materials. *Geoderma* **2002**, *108*, 237–257. [CrossRef]

34. Prietzel, J.; Thieme, J.; Eusterhues, K.; Eichert, D. Iron speciation in soils and soil aggregates by synchrotron-based X-ray microspectroscopy (xanes, mu-xanes). *Eur. J. Soil Sci.* **2007**, *58*, 1027–1041. [CrossRef]

35. Formenti, P.; Caquineau, S.; Chevaillier, S.; Klaver, A.; Desboeufs, K.; Rajot, J.; Belin, S.; Briois, V. Dominance of goethite over hematite in iron oxides of mineral dust from western Africa: Quantitative partitioning by X-ray absorption spectroscopy. *J. Geophys. Res.-Atmos.* **2014**, *119*, 12740–12754. [CrossRef]

36. Wilke, M.; Farges, F.; Petit, P.; Brown, G.; Martin, F. Oxidation state and coordination of Fe in minerals: An fek-xanes spectroscopic study. *Am. Mineral.* **2001**, *86*, 714–730. [CrossRef]

37. Berry, A.; O'Neill, H.; Jayasuriya, K.; Campbell, S.; Foran, G. Xanes calibrations for the oxidation state of iron in a silicate glass. *Am. Mineral.* **2003**, *88*, 967–977. [CrossRef]

38. Balzarotti, A.; Bianconi, A.; Burattini, E.; Grandolfo, M.; Habel, R.; Piacentini, M. Core transitions from the al 2p level in amorphous and crystalline Al_2O_3. *Phys. Status Solidi (b)* **1974**, *63*, 77–87. [CrossRef]

39. Belli, M.; Scafati, A.; Bianconi, A.; Mobilio, S.; Palladino, L.; Reale, A.; Burattini, E. X-ray absorption near edge structures (XANES) in simple and complex Mn compounds. *Solid State Commun.* **1980**, *35*, 355–361. [CrossRef]

40. Benfatto, M.; Natoli, C.R.; Bianconi, A.; Garcia, J.; Marcelli, A.; Fanfoni, M.; Davoli, I. Multiple-scattering regime and higher-order correlations in X-ray-absorption spectra of liquid solutions. *Phys. Rev. B* **1986**, *34*, 5774–5781. [CrossRef]

41. Koningsberger, D.C.; Prins, R. (Eds.) *X-ray Absorption: Principles, Applications, Techniques of EXAFS, SEXAFS, and XANES Chemical Analysis*; Wiley: New York, NY, USA, 1988; Volume 2, pp. 1–673.

42. Bianconi, A.; Fritsch, E.; Calas, G.; Petiau, J. X-ray-absorption near-edge structure of 3d transition elements in tetrahedral coordination: The effect of bond-length variation. *Phys. Rev. B Condense. Matter* **1985**, *32*, 4292–4295. [CrossRef]

43. Dent, A.J.; Cibin, G.; Ramos, S.; Smith, A.D.; Scott, S.M.; Varandas, L.; Pearson, M.R.; Krumpa, N.A.; Jones, C.P.; Robbins, P.E. B18: A Core XAS Spectroscopy Beamline for Diamond. *J. Phys. Conf. Ser.* **2009**, *190*, 012039. [CrossRef]

44. Marcelli, A.; Hampai, D.; Giannone, F.; Sala, M.; Maggi, V.; Marino, F.; Pignotti, S.; Cibin, G. Xrf-xanes characterization of deep ice core insoluble dust. *J. Anal. At. Spectrom.* **2012**, *27*, 33–37. [CrossRef]

45. Cibin, G.; Marcelli, A.; Maggi, V.; Sala, M.; Marino, F.; Delmonte, B.; Albani, S.; Pignotti, S. First combined total reflection X-ray fluorescence and grazing incidence X-ray absorption spectroscopy characterization of aeolian dust archived in Antarctica and alpine deep ice cores. *Spectrochim. Acta Part B-At. Spectrosc.* **2008**, *63*, 1503–1510. [CrossRef]

46. Davey, B.; Russell, J.; Wilson, M. Iron-oxide and clay-minerals and their relation to colors of red and yellow podzolic soils near Sydney, Australia. *Geoderma* **1975**, *14*, 125–138. [CrossRef]

47. Singh, B.; Gilkes, R. Properties and distribution of iron-oxides and their association with minor elements in the soils of south-western Australia. *J. Soil Sci.* **1992**, *43*, 77–98. [CrossRef]

48. Tombolini, F.; Brigatti, M.F.; Marcelli, A.; Cibin, G.; Mottana, A.; Giuli, G. Local and average Fe distribution in trioctahedral micas: Analysis of Fe K-edge XANES spectra in the phlogopite–annite and phlogopite tetraferriphlogopite joins on the basis of single-crystal XRD refinements. *Eur. J. Mineral.* **2002**, *14*, 1075–1085. [CrossRef]

49. Delmonte, B.; Petit, J.; Basile-Doelsch, I.; Lipenkov, V.; Maggi, V. First characterization and dating of east Antarctic bedrock inclusions from Subglacial lake vostok accreted ice. *Environ. Chem.* **2004**, *1*, 90–94. [CrossRef]

50. Shoenfelt, E.; Sun, J.; Winckler, G.; Kaplan, M.; Borunda, A.; Farrell, K.; Moreno, P.; Gaiero, D.; Recasens, C.; Sambrotto, R.; et al. High particulate iron(ii) content in glacially sourced dusts enhances productivity of a model diatom. *Sci. Adv.* **2017**, *3*, e1700314. [CrossRef]

51. Delmonte, B.; Andersson, P.; Schoberg, H.; Hansson, M.; Petit, J.; Delmas, R.; Gaiero, D.; Maggi, V.; Frezzotti, M. Geographic provenance of aeolian dust in east Antarctica during Pleistocene glaciations: Preliminary results from talos dome and comparison with east Antarctic and new Andean ice core data. *Quat. Sci. Rev.* **2010**, *29*, 256–264. [CrossRef]

52. Baccolo, G.; Delmonte, B.; Albani, S.; Baroni, C.; Cibin, G.; Frezzotti, M.; Hampai, D.; Marcelli, A.; Revel, M.; Salvatore, M.C.; et al. Regionalization of the atmospheric dust cycle on the periphery of the East Antarctic ice sheet since the last glacial maximum. *Geochem. Geophys. Geosyst.* **2018**. [CrossRef]

53. De Angelis, M.; Tison, J.L.; Morel-Fourcade, M.C.; Susini, J. Micro-investigation of EPICA Dome C bottom ice: Evidence of long term in situ processes involving acid-salt interactions, mineral dust, and organic matter. *Quat. Sci. Rev.* **2013**, *78*, 248–265. [CrossRef]

54. Baccolo, G.; Cibin, G.; Delmonte, B.; Hampai, D.; Marcelli, A.; Di Stefano, E.; Macis, S.; Maggi, V. The Contribution of Synchrotron Light for the Characterization of Atmospheric Mineral Dust in Deep Ice Cores: Preliminary Results from the Talos Dome Ice Core (East Antarctica). *Condens. Matter* **2018**, *3*, 25. [CrossRef]

55. Galdenzi, F.; Della Ventura, G.; Cibin, G.; Macis, S.; Marcelli, A. Accurate Fe^{3+}/Fe_{tot} ratio from XAS spectra at the Fe K-edge. *Radiat. Phys. Chem.* **2018**, in press.

Article

The Contribution of Synchrotron Light for the Characterization of Atmospheric Mineral Dust in Deep Ice Cores: Preliminary Results from the Talos Dome Ice Core (East Antarctica)

Giovanni Baccolo [1,2,*] , Giannantonio Cibin [3], Barbara Delmonte [1], Dariush Hampai [4] , Augusto Marcelli [4,5] , Elena Di Stefano [1,2,6], Salvatore Macis [4,7] and Valter Maggi [1,2]

[1] Environmental and Earth Sciences Department, University Milano-Bicocca, 20126 Milano, Italy; barbara.delmonte@unimib.it (B.D.); elena.distefano@mib.infn.it (E.D.S.); valter.maggi@unimib.it (V.M.)
[2] Milano-Bicocca Section, Istituto Nazionale di Fisica Nucleare, 20126 Milano, Italy
[3] Diamond Light Source, Harwell Science and Innovation Campus, Didcot OX11 0DE, UK; giannantonio.cibin@diamond.ac.uk
[4] Laboratori Nazionali di Frascati, Istituto Nazionale di Fisica Nucleare, 00044 Frascati, Italy; dariush.hampai@lnf.infn.it (D.H.); augusto.marcelli@lnf.infn.it (A.M.); salvatore.macis91@gmail.com (S.M.)
[5] RICMASS, Rome International Center for Materials Science Superstripes, 00185 Roma, Italy
[6] Department of Physical, Earth and Environmental Sciences, University of Siena, 53100 Siena, Italy
[7] Department of Mathematics and Physics, University Roma-Tor Vergata, 00133 Rome, Italy
* Correspondence: giovanni.baccolo@mib.infn.it; Tel.: +39-02-6448-2435

Received: 4 July 2018; Accepted: 26 August 2018; Published: 28 August 2018

Abstract: The possibility of finding a stratigraphically intact ice sequence with a potential basal age exceeding one million years in Antarctica is giving renewed interest to deep ice coring operations. But the older and deeper the ice, the more impactful are the post-depositional processes that alter and modify the information entrapped within ice layers. Understanding in situ post-depositional processes occurring in the deeper part of ice cores is essential to comprehend how the climatic signals are preserved in deep ice, and consequently how to construct the paleoclimatic records. New techniques and new interpretative tools are required for these purposes. In this respect, the application of synchrotron light to microgram-sized atmospheric dust samples extracted from deep ice cores is extremely promising. We present here preliminary results on two sets of samples retrieved from the Talos Dome Antarctic ice core. A first set is composed by samples from the stratigraphically intact upper part of the core, the second by samples retrieved from the deeper part of the core that is still undated. Two techniques based on synchrotron light allowed us to characterize the dust samples, showing that mineral particles entrapped in the deepest ice layers display altered elemental composition and anomalies concerning iron geochemistry, besides being affected by inter-particle aggregation.

Keywords: atmospheric mineral dust; ice core; Antarctica; paleoclimate; synchrotron radiation; X-ray absorption near edge spectroscopy; X-ray fluorescence; iron geochemistry

---

1. Introduction

Atmospheric mineral dust is a key component of the Earth climate system. Climate influences the production, transport, and deposition of dust. Conversely, dust affects climate, with direct and indirect effects related to the radiative properties of the atmosphere and of the surfaces where it is deposited, to cloud physics, and to biogeochemistry [1–3]. Paleoclimate archives were essential to reconstruct the past interactions between climate and the global dust cycle. Ice cores showed that a close connection exists between them at different spatial and temporal scales: From fast climatic oscillations to the longer glacial/interglacial transitions [4–6]. At present, EPICA Dome C is the Antarctic ice core that allowed for the longest and most detailed climatic reconstruction of the last 800,000 years [7,8], including an accurate record of dust deposition during the last eight climatic cycles [4].

The need for retrieving a deep ice core extending beyond EPICA Dome C and encompassing the so-called Mid Pleistocene revolution is definitely challenging. A key aspect related to signal preservation in the bottom part of the core depends on our understanding of in situ post-depositional processes potentially affecting the original climate related signals [9–11]. Disentangling the latter from the true climatic signals embedded in deep and ancient ice will be stimulating for the field of ice core science, but also for analytical chemistry and ice micro-physics. The processes usually quoted as the most important ones, in relation to the occurrence of post depositional processes in deep ice, are: The evolution of ice crystal fabric, the diffusion and/or relocation of soluble and insoluble impurities, and the interaction with bedrock [9,10,12].

Mineral dust enclosed in the deeper part of ice cores—where the temperature gradient increases and the ice temperature is close to the pressure melting point—is known to be affected by relocation processes. Impurity relocation occurs because of the exclusion of solid impurities from the ice lattice during ice grain recrystallization. A direct consequence of this process is the enrichment of particles in the unfrozen liquid film that separates grain boundaries, and the formation of dust aggregates of different sizes, as first noticed in the EPICA Dome C ice cores below about 2900 m depth [11]. Additional processes potentially affecting the dust record in the bottom part of deep ice cores are related to the presence of inclusions from the bedrock, which makes atmospheric dust particles more difficult to recognize and analyze [10].

In this work, we present preliminary results about the atmospheric mineral dust record from the deep part of the Talos Dome ice core (TALDICE). We show that in addition to the aggregation of mineral particles, the dust entrapped in deep ice is also affected by evident compositional changes, in particular concerning iron geochemistry. These findings are coherent with early observations on the deeper part of the EPICA Dome C ice core, where evidence of chemical alteration of insoluble particles were shown by De Angelis and coauthors [10]. Despite being preliminary, our results show that synchrotron radiation-based techniques are extremely promising for ice core science.

2. Materials and Methods

2.1. The Talos Dome Ice Core

The Talos Dome ice core—also known as TALDICE—was drilled from the homonymous ice dome located at the NE periphery of the East Antarctic Ice Sheet (EAIS), close to Victoria Land and the Ross Sea [13]. It is 1620 m long and can be divided into two parts. The upper part extends from the surface to a depth of 1440 m, it was recognized to be stratigraphically intact and corresponds to the last 150 kyr [14]. The lower section ranges from 1440 to the end of the core; its integrity has not been assessed because of relevant concerns about the influence of the irregular topography of the underlying bedrock that characterizes the Talos Dome area [15]. To the aims of this study, two sets of samples were considered, consisting of 4 samples each: The first one (from TD1 to TD4) is composed by samples retrieved from the upper part of the core that is climatically intact [16], the second group (from TD5 to TD8) is composed by samples from the deeper part that corresponds to the undated section (below 1440 m). Samples were prepared using ice strips dedicated to dust analysis. Ice core strips were decontaminated and treated for the successive analyses in accordance to a well-established procedure [17]. After melting, decontaminated meltwater was divided into two aliquots, one for Coulter counter analysis, one for synchrotron light-based measurements. Details about the samples are found in Table 1.

Table 1. Details about the samples from the Talos Dome ice core considered in this work. Depth refers to the depth along the core with respect to ground level; ice age (expressed in terms of thousands of years before 1950, i.e., kyr BP) was retrieved from the AICC2012 chronology developed for some Antarctic ice cores [15,16]; the dust concentration was determined through Coulter counter analysis. MIS2 corresponds to Marine Isotopic Stage 2: It is the coldest and final phase of the last glacial period, it occurred between 30 and 18 kyr BP.

Sample	Depth (m)	Ice Age (kyr BP)	Climatic Period	Dust Conc. ($ng_{dust}·g^{-1}_{ice}$)
TD1	185	2.0	Holocene	27 ± 5
TD2	241	2.8	Holocene	27 ± 2
TD3	857	20.2	MIS2	479 ± 5
TD4	873	21.8	MIS2	512 ± 6
TD5	1530	Unkn.	Deep Part	230 ± 10
TD6	1554	Unkn.	Deep Part	133 ± 8
TD7	1595	Unkn.	Deep Part	25.2 ± 0.7
TD8	1613	Unkn.	Deep Part	35 ± 3

2.2. Coulter Counter

The Coulter counter technique is well-established dealing with ice cores [18]. It allows for the detection of micrometric insoluble mineral particles suspended within a liquid water solution. It employs the impedance generated by the flowing of a solid particle through a narrow orifice where an electric voltage is applied. Further details about the technique and about the procedure are found in previous publications [18]. For this work, a 30 μm orifice was used and particles between 0.6 and 18 μm (divided into 400 channels) were detected.

2.3. Synchrotron Light Measurements

Two X-ray spectroscopy techniques based on synchrotron radiation were applied to the atmospheric dust samples from the Talos Dome ice core: 1 X-ray fluorescence (XRF); and 2 X-ray absorption near edge structure spectroscopy (XANES). These tools provide complementary and diverse pieces of information [19]. Through XRF, it was possible to assess the elemental composition of the dust particles extracted from the ice core strips; XANES, on the other hand, allowed gathering detailed information about selected elements. In the case of XRF, attention was given to major elements, while in the case of XANES we focused on iron. Both major elements and iron are geochemically relevant. Major elements, i.e., the ones whose oxide mass fraction exceeds 0.1% of average crustal mass (Na, Mg, Al, Si, K, Ca, Ti, Mn, and Fe), constitute more than 99% of Earth crust [20]. Because of their abundance, they are relatively easy to determine, but at the same time they allow one to depict a coherent geochemical signature of the considered crustal material. Indeed, they were already used to characterize atmospheric mineral dust from ice cores [21]. Among major elements, Fe is one of the most interesting when considering the interactions between biogeochemistry, dust, and climate. This is because atmospheric dust plays a role in supplying iron to the oceans, where this element is a limiting factor for oceanic bio-productivity [2,22]. Ice cores are important in this context, since they allow reconstructing the variability of these processes on the glacial-interglacial timescale and assessing the role of dust in the global carbon cycle [23,24].

Samples dedicated to synchrotron light measurements were prepared through filtration. After decontamination and melting, mineral dust particles were extracted from meltwater with a filtration system coupled to clean (rinsed with a 3% HNO_3 solution for one month) polycarbonate membranes (0.45 μm pore size) which proved to be sufficiently clean for our purposes [25]. A comparison between the XRF signal related to blank membranes and our samples is shown in Figure 1. It can be appreciated that despite samples consisting a few μg of dust, their signal is at least one order of magnitude higher than the blank one. This means that the contamination in the blank membranes is in the order of few tens of pg for each element. A variable amount of meltwater was filtered for each sample, so as to deposit at least 2–3 µg of dust on each membrane. The filtration was carried out using a micro-pipette, transferring meltwater one mL at a time on the membrane mounted on the filtration system. This step is important since the accumulation of dust particles in the smallest possible area greatly improves the quality of analytical results [26]. Before transferring the meltwater from the original containers to the filtration system, the containers were shaken, so as to mobilize the particles eventually deposited on the bottom of the container. When all the meltwater was transferred, a few mL of ultrapure water is used to rinse the inner walls of the container tubes to catch the mineral particles that could be attached to them. Rinsing water was then passed to the filtration system using the pipette. This procedure was repeated three times for each sample and also for the preparation of the blanks. After filtration, the membranes were mounted on clean PTFE sample holders and successively analyzed at the B18 beamline of the Diamond Light Source [27].

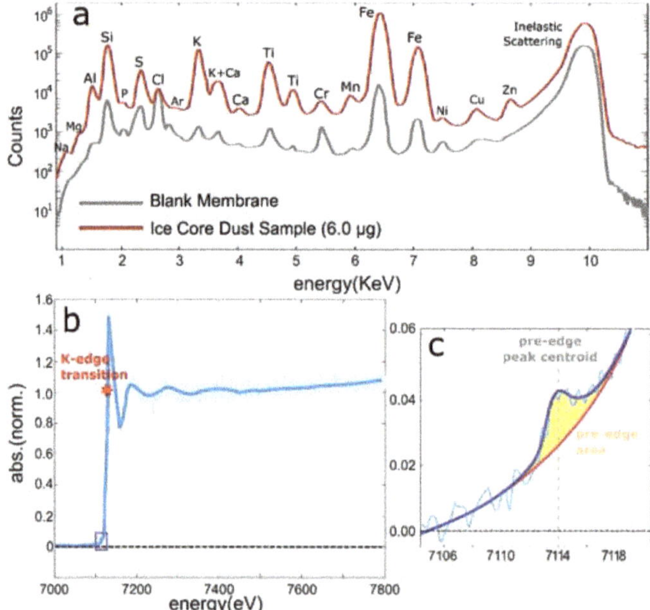

Figure 1. Examples of the results obtained through the application of synchrotron radiation to ice core atmospheric dust samples. Panel (**a**): An X-ray fluorescence (XRF) spectrum; it is possible to appreciate the difference between a blank membrane and an average dust sample (total mass deposited on the filter 6 µg). Panels (**b**,**c**) show an X-ray absorption near edge structure spectroscopy (XANES) spectrum related to the Fe K adsorption edge. The red star refers to the position of the edge, identified using the second derivative. In panel (**c**), an enlargement of the pre-edge region is shown, including a fit of the background (red curve) and of the pre-edge peak (dark blue curve). The two parameters used for the pre-edge analysis are highlighted: The integral area of the pre-edge structure (yellow area) and its centroid energy (vertical dashed line).

A silicon drift detector was used for both XRF and XANES measurements. A complete description of the instrumental setup can be found here [28]. XRF spectra were collected with a 10 keV incident beam and 600 s acquisition time. XANES was performed on Fe K-edge considering the energy interval between 6900 and 7800 eV and steps of 0.15 eV between each acquisition. Energy was calibrated with repeated measurements of the absorption edge of a metallic Fe foil (7122 eV), additional details are given in previous works [29,30], while explanatory results are presented in Figure 1.

3. Results and Discussion

3.1. Grain Size Distributions

The two sets of atmospheric mineral dust samples extracted from the Talos Dome ice core present evident differences in relation to all the variables and parameters considered in this study: Grain size distribution, elemental composition, and iron properties. In Figure 2, it is possible to appreciate the differences characterizing grain size distributions. The first two distributions refer to two samples from the first group considered here, i.e., the one composed by samples obtained from the stratigraphically intact section of the Talos Dome ice core. They display clear differences owing to the different climatic periods during which they were deposited. The sample from the Holocene (Figure 2a) presents low dust concentration (27 $ng_{dust} \cdot g^{-1}_{ice}$) and a poor size selection; on the contrary, the sample from MIS2 is more concentrated (479 $ng_{dust} \cdot g^{-1}_{ice}$) and displays a distribution that is easily fitted by a log-normal equation, ascribable to a prolonged atmospheric transport. These features are well-known for the Talos Dome ice core: They were related to different dust depositional regimes characterizing the site during glacial and interglacial periods [31–33]. What observed in Figure 2c is different with respect to the previous cases. The third distribution concerns a sample from the deep undated part of the ice core. It shows a coarse mode (4.4 µm), with limited presence of fine particles. Such features are never encountered when dealing with more superficial sections of Antarctic ice cores, considering either interglacial periods, either glacial ones [34]. The only process that can explain similar features is particle aggregation. Its occurrence was already observed in deep polar ice cores and it was explained as the effect of ice grain growth on mineral dust particles [10,11]. But if such effects on grain size distribution are partially known, much less is understood when pointing attention to dust compositional changes in deep ice.

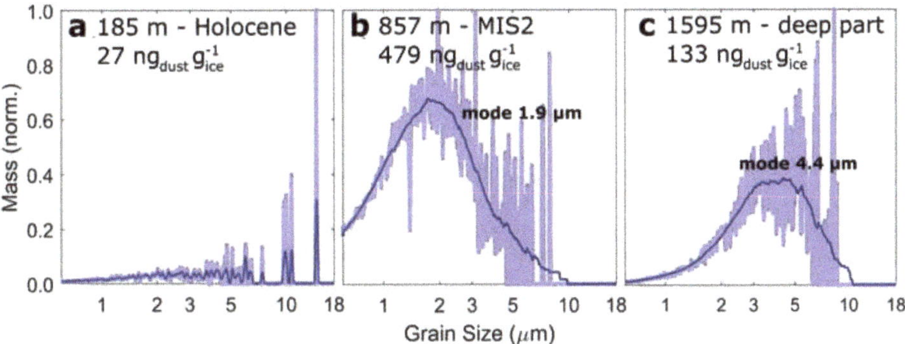

Figure 2. Examples of size distributions determined through Coulter counter of the mineral particles found in ice sections from the Talos Dome ice core. For each sample the depth along the core, the climatic period and the absolute dust concentration are reported. From the left to the right. (**a**) A sample from the Holocene, showing poor size selection and a large relative contribution from coarse particles; (**b**) A sample from Marine Isotopic Stage 2 (MIS2, one of the three stages of the last glacial period), in this case size selection is evident and points to a prolonged atmospheric transport; (**c**) A sample from the deep undated part of the Talos Dome ice core; the coarse modal value and the limited presence of fine particles point to a significant aggregation of mineral particles.

3.2. Dust Composition

Data about the major element composition of the Talos Dome ice core samples are fully reported in Table 2, while they are graphically presented in Figure 3. It can be easily appreciated that the samples gathered from the climatically intact sections of the core show different compositional features with respect to the ones from the deeper part. The composition of the shallower samples is similar to the

Upper Continental Crust reference (UCC, [35]), confirming the crustal origin for the insoluble particles deposited on polar ice sheets. The only significant difference concerns CaO: The average CaO content in TD shallow samples is 1.7%, in accordance to what observed at another Antarctic site [21], while the value for average UCC is 3.6%. Calcium carbonate accounts for a significant fraction of Ca in global crust sediments, but dust transported toward Antarctica from the continental areas of the Southern hemisphere is depleted in carbonate as a consequence of acid-base reactions occurring in the atmosphere during the transport [36,37]. Additional differences are appreciated if comparing the four samples constituting this group, for example the variable amount of Al and Si oxides. Such variability is not unexpected, since these samples are partly from the current interglacial period (the Holocene) and partly from the last glacial maximum. It is known that the dust sources responsible for the deposition of mineral particles at Talos Dome were different in the two periods, with effects on dust composition [28,38].

Table 2. Major element composition of the samples. Since the total mass of major element oxide represents more than 99% of average Earth crust composition [20], the total sum was closed to 100%. For each one of the two sample groups (respectively shallow and deep samples) the average composition is also reported (standard deviations in brackets). Upper continental crust (UCC) reference from Rudnick & Gao [35].

Sample	Na_2O%	MgO%	Al_2O_3%	SiO_2%	K_2O%	CaO%	TiO_2%	MnO%	Fe_2O_3%
TD1	2.0	1.5	18.6	67.9	1.9	1.3	0.8	0.05	6.1
TD2	5.5	2.1	16.1	64.2	1.4	2.5	1.1	0.07	6.9
TD3	2.1	6.0	21.0	61.3	1.7	1.5	1.0	0.07	5.3
TD4	2.4	2.1	22.2	60.2	2.0	1.6	1.5	0.11	7.8
Average shallow	3.0	2.9	19.5	63.4	1.8	1.7	1.1	0.07	6.5
samples	(1.7)	(2.1)	(2.7)	(3.4)	(0.3)	(0.6)	(0.3)	(0.03)	(1.1)
TD5	2.6	0.9	21.3	65.1	2.7	0.3	0.9	0.04	6.0
TD6	0.8	0.6	15.7	72.7	2.4	1.1	1.2	0.04	5.6
TD7	0.9	0.2	7.7	84.0	1.4	0.2	0.8	0.02	4.7
TD8	2.0	0.1	7.4	80.0	1.9	0.2	1.0	0.03	7.3
Average deep	1.6	0.5	13.0	75.4	2.1	0.4	1.0	0.03	5.9
samples	(0.9)	(0.4)	(6.7)	(8.3)	(0.6)	(0.4)	(0.2)	(0.01)	(1.1)
UCC reference	3.3	2.5	15.4	66.7	2.8	3.6	0.6	0.1	5.0

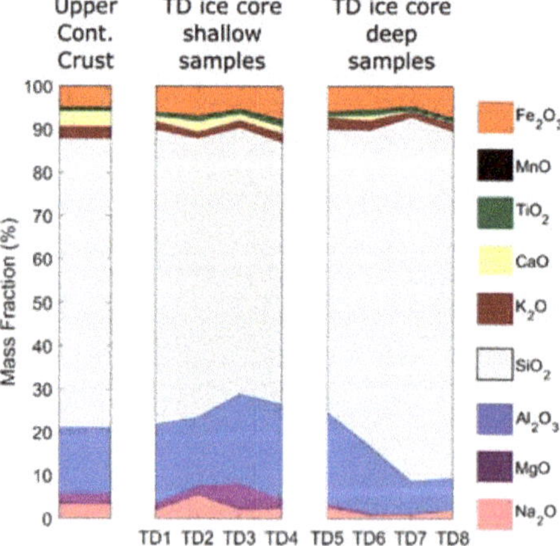

Figure 3. Major element composition of the mineral dust samples extracted from the Talos Dome ice core. From left to right: The UCC reference [35], samples from the shallower section of the Talos Dome ice core, samples from the deeper one.

A different scenario is found when focusing on the samples from the deeper part of the Talos Dome ice core. In this case, their average composition shows clear dissimilarities with respect to the UCC reference. This is particularly true for Mg and Ca oxides. Both are almost absent in the deep samples, with average concentrations of 0.5 and 0.4%, representing respectively 1/9 and 1/5 of UCC composition. These two elements were already recognized as some of the most affected by post-depositional processes in deep ice cores. An increase of the concentration of dissolved Ca^{2+} and Mg^{2+} was observed in the deepest sections of the EPICA Dome C ice core, in association with anomalies concerning sulfates and other dissolved species [10,33]. Considering this context, it is possible to hypothesize that in deep ice mineral particles are subjected to significative chemical transformations. The depletion of CaO and MgO in mineral dust from the deeper sections of the Talos Dome ice core and the rise of the dissolved concentration of Ca^{2+} and Mg^{2+} in other deep Antarctic ice cores could be explained by a progressive dissolution of the carbonate mineral fraction originally present within the mineral dust particles. In deep ice, the concurrent presence of liquid veins [39] and dust aggregates at grain boundaries [9] and of an acid environment [36], could enhance such dissolution processes. CaO and MgO are the two oxides for which these effects are more evident, but it seems that for Al_2O_3, NaO, and Fe_2O_3 they could also be active, given their significant depletion in deep samples. As a complementary evidence, the SiO_2 fraction of dust particles is larger in the deep part of the Talos Dome ice core than in the shallow one (on average 75% vs. 63%). This is a consequence of the geochemical stability of SiO_2: If the more labile mineral fractions are dissolved, the stable ones increase their relative abundance.

3.3. Iron Geochemistry

In Figure 4, the data from Fe-XANES are presented. Figure 4a refers to the absorption spectra and show the transition of Fe K-edge in all the samples. The data used to interpret iron geochemistry were obtained from the analysis of these spectra. Attention was given to different spectral regions: The pre-edge one (Figure 4b) and the position of the main energetic transition edge (Figure 4c). Both are sensitive to iron speciation, oxidation, and coordination states [40]. Unfortunately, it was not possible to gather data from mineralogical standards and for this reason we cannot quantitatively assess the mineralogical assemblages that could resemble our results. Such an analysis will be discussed in the near future, but a qualitative look at the spectra is sufficient to reveal that the spectra show some differences. Despite this limitation, it was still possible to extract information about the oxidation and coordination states of our samples, as shown in Figure 4b,c.

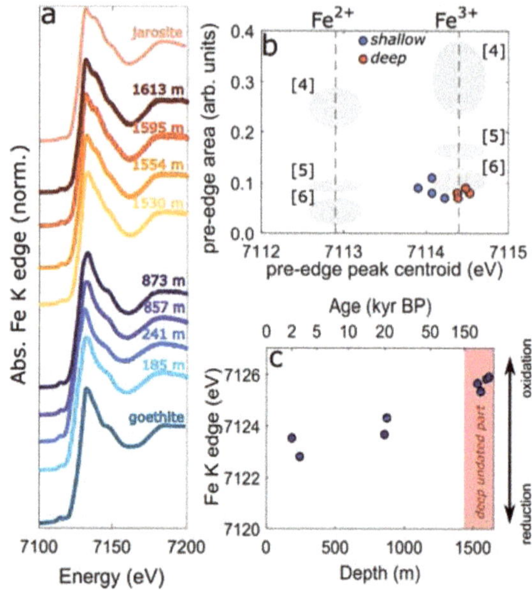

Figure 4. Results from XANES Fe spectroscopy. Panel (**a**): X-ray absorption spectra at the Fe K-edge. For each spectrum the depth of the considered sample along the Talos Dome ice core is reported. Blue samples are from the intact part of the core. Yellow/orange ones are from the deeper undated part below 1440 m. In addition, two spectra referred to mineral standards are shown: Goethite [41] and jarosite [42], which were retrieved from literature and have to be considered as merely indicative. Panel (**b**): Pre-edge analysis of the absorption spectra (see also the caption of Figure 1). For each sample, the following variables were considered: The integral area of the pre-edge peak and its centroid energy. Data uncertainties are comparable to the size of the circles. The two-axes domain is presented in accordance to a previously proposed scheme [43,44], where characteristic fields are recognized in relation to iron oxidation (ferrous Fe^{2+} and ferric Fe^{3+}) and coordination (tetrahedral [4], trigonal bipyramidal [5], and octahedral [6], please note that in this case the notation [...] doesn't refer to the cited bibliography but to atomic coordination numbers) states. Panel (**c**): A comparison between the Fe K-edge energy observed for each sample and their depth along the Talos Dome ice core (ice age is also reported [16]).

Data about the pre-edge region of the spectra—i.e., the integral area of the pre-edge peak and its energy position—were presented using the scheme proposed by Wilke and colleagues [43] and Giuli and coauthors [44]. In that scheme, the two-dimensional space is divided into different domains reflecting the oxidation and coordination states of Fe. Not unexpectedly, all the samples are found near the field of [6] Fe^{3+} ([6] doesn't refer to bibliography, but to iron coordination number, see the caption of Figure 4 for further details), that is iron in its ferric state with an octahedral geometry. All the most common iron oxides that are found in superficial environments belong to this field, for example goethite, hematite, and ferrihydrite [43], confirming that iron transported in association with atmospheric dust is mainly related to these minerals [45]. Despite all the samples being located near the [6] Fe^{3+} field, a difference is observed between shallow and deep samples (respectively blue and red dots in Figure 4b). The first ones show a pre-edge energy which is in accordance to a mix of 70% Fe^{3+} and 30% Fe^{2+}, on the contrary the signature of the deep samples reveals a 100% Fe^{3+} contribution. The same is supported by Figure 4c, where the position of the K-edge is compared to the depth of the samples along the ice core. The deeper the samples, the more oxidized the iron. This is an additional clue about the effects of post-depositional processes on mineral dust composition in deep ice. Indeed, deep ice sheets are not anoxic environments, the presence of liquid water veins [39] and dissolved oxygen [46] could enhance not only the dissolution of specific mineral phases but also their oxidation. XANES spectra of the deeper samples present a spectral feature at about 7140 eV, which the shallower samples don't display (see Figure 4a). This is an additional evidence for the occurrence of post-depositional processes, since that feature is typical only in some Fe-sulfates, such as jarosite [42] (Figure 4a). The latter, which presents iron in its [6] Fe^{3+} state, is known to form because of the oxidation and alteration of Fe minerals in acid environments where sulfur-rich fluids and water are present [47]. Sulfates are known for being subjected to heavy post-depositional processes in deep ice sheets [10,38]. It was already proposed that in deep ice sulfuric acid could react with mineral phases, enhancing the in-situ production of sulfate salts, as for example $MgSO_4$ [38]. Combining our preliminary evidences about iron speciation and the direct observation of localized sulfate deposits on mineral dust particles in deep ice [10], it is reasonable to hypothesize that the iron fraction of mineral dust entrapped in deep ice is affected by relevant post-depositional processes. Future studies focused on the reconstruction of ice core records about the oxidation of iron in the past climatic periods and on its role in iron biogeochemistry should take these results into account [24,48].

4. Conclusions and Perspectives

Although being preliminary, the current work clearly shows the great potential of synchrotron light for the characterization of mineral dust extracted from ice core samples. Combining XRF, XANES, and traditional grain size analysis, it was possible to appreciate significant differences between mineral dust samples retrieved from the shallow part of the Antarctic Talos Dome ice core and from the deep one. It was already known that in deep ice, dust is affected by aggregation and re-location; what is new is that its chemical composition is also influenced by ice depth. Dissolution and oxidation seem to be the most important processes in this context, but further efforts are needed to better comprehend them and to clearly assess the role played by post-depositional process in altering ice core records of atmospheric mineral dust. This is a preliminary work, focused on few samples from a single ice core and on a single K-edge transition. Its main purpose is to demonstrate the potential of such techniques in relation to the future deep ice core drilling projects, where the assessment of the integrity of the climatic signals embedded within the ice stratigraphy will became more and more important.

Author Contributions: A.M. and V.M. conceived the idea of this work. A.M. and G.C. designed the experiments; all the authors prepared the samples and performed the experiments; G.C., D.H. and G.B. analyzed the data; G.B. wrote the paper and interpreted the results.

Acknowledgments: The Talos Dome Ice core Project (TALDICE), a joint European programme, is funded by national contributions from Italy, France, Germany, Switzerland and the United Kingdom. Primary logistical support was provided by PNRA at Talos Dome. This is TALDICE publication no 50. This study has been partially supported by DARA (Dipartimento per gli Affari Regionali e Autonomie of Italian Presidenza del Consiglio dei Ministri) in the framework of the MIAMI (Monitoraggio dell'Inquinamento Atmosferico della Montagna Italiana) project. Synchrotron radiation measurements were carried out in the framework of Proposals NT1984 at the Diamond Light Source.

References

1. Arimoto, R. Eolian dust and climate: Relationships to sources, tropospheric chemistry, transport and deposition. *Earth Sci. Rev.* **2001**, *54*, 29–42. [CrossRef]
2. Jickells, T.D.; An, Z.S.; Andersen, K.K.; Baker, A.R.; Bergametti, G.; Brooks, N.; Cao, J.J.; Boyd, P.W.; Duce, R.A.; Hunter, K.A.; et al. Global iron connections between desert dust, ocean biogeochemistry and climate. *Science* **2005**, *308*, 67–71. [CrossRef] [PubMed]
3. Mahei, D.A.; Prospero, J.M.; Mackie, D.; Gaiero, D.; Hesse, P.P.; Balkanski, Y. Global connections between aeolian dust, climate and ocean biogeochemistry at the present day and at the last glacial maximum. *Earth-Sci. Rev.* **2010**, *99*, 61–97. [CrossRef]

4. Lambert, F.; Delmonte, B.; Petit, J.R.; Bigler, M.; Kaufmann, P.R.; Hutterli, M.A.; Stocker, T.F.; Ruth, U.; Steffensen, J.P.; Maggi, V. Dust-climate couplings over the past 800.000 years from the EPICA Dome C ice core. *Nature* **2008**, *452*, 616–619. [CrossRef] [PubMed]

5. Fischer, H.; Fundel, F.; Ruth, U.; Twarloh, B.; Wegner, A.; Udisti, R.; Becagli, S.; Castellano, E.; Morganti, A.; Severi, M.; et al. Reconstruction of millenial changes in dust emission, transport and regional sea ice coverage using the depp EPICA ice cores from the Atlantic and Inidian Ocean secont of Antarctica. *Earth Planet. Sci. Lett.* **2007**, *260*, 340–354. [CrossRef]

6. Wegner, A.; Fischer, H.; Delmonte, B.; Petit, J.R.; Erhardt, T.; Ruth, U.; Svensson, A.; Vinther, B.; Miller, H. The role of seasonality of mineral dust concentration and size on glacial/interglacial dust changes in the EPICA Dronning Maud Land ice core. *J. Geophys. Res.* **2015**, *120*. [CrossRef]

7. Epica Community Members. Eight glacial cycles from an Antarctic ice core. *Nature* **2004**, *429*, 623–628. [CrossRef] [PubMed]

8. Jouzel, J.; Masson-Delmotte, V.; Cattani, O.; Dreyfus, G.; Falourd, S.; Hoffmann, G.; Minster, B.; Nouet, J.; Barnola, J.M.; Chappellaz, J.; et al. Orbital and millennial Antarctic climate variability over the past 800,000 years. *Nature* **2007**, *317*, 793–796. [CrossRef] [PubMed]

9. Faria, S.H.; Freitag, J.; Kipstul, S. Polar ice structure and the integrity of ice-core paleoclimate records. *Quat. Sci. Rev.* **2010**, *29*, 338–351. [CrossRef]

10. De Angelis, M.; Tison, J.L.; Morel-Fourcade, M.C.; Susini, J. Micro-investigation of EPICA Dome C bottom ice: Evidence of long term in situ processes involving acid-salt interactions, mineral dust, and organic matter. *Quat. Sci. Rev.* **2013**, *78*, 248–265. [CrossRef]

11. Tison, J.L.; de Angelis, M.; Littot, G.; Wolff, E.; Fischer, H.; Hansson, M.; Bigler, M.; Udisti, R.; Wegner, A.; Jouzel, J.; et al. Retrieving the paleoclimatic signal from the deeper part of the EPICA Dome C ice core. *Cryosphere* **2015**, *9*, 1633–1648. [CrossRef]

12. Goossens, T.; Sapart, C.J.; Dahl-Jensen, D.; Popp, T.; El Amri, S.; Tison, J.L. A comprehensive interpretation of the NEEM basal ice build-up using a multi-parametric approach. *Cryosphere* **2016**, *10*, 553–567. [CrossRef]

13. Frezzotti, M.; Bitelli, G.; De Michelis, P.; Deponti, A.; Forieri, A.; Gandolfi, S.; Maggi, V.; Mancini, F.; Remy, F.; Tabacco, I.; et al. Geophysical survey at Talos Dome, East Antarctica: The search for a new deep-drilling site. *Ann. Glaciol.* **2004**, *39*, 423–432. [CrossRef]

14. Stenni, B.; Buiron, D.; Frezzotti, M.; Albani, S.; Barbante, C.; Bard, E.; Barnola, J.M.; Baroni, M.; Baumgartner, M.; Bonazza, M.; et al. Expression of the bipolar see-saw in Antarctic climate record during the last deglaciation. *Nat. Geosci.* **2011**, *4*, 46–49. [CrossRef]

15. Bazin, L.; Landais, A.; Lemieux-Dudon, B.; Toyé Mahamado Kele, H.; Veres, D.; Parrenin, F.; Martinerie, P.; Ritz, C.; Capron, E.; Lipenkov, V.; et al. An optimized multi-proxy, multi-site Antarctic ice and gas orbital chronology (AICC2012): 120–800 ka. *Clim. Past* **2013**, *9*, 1715–1721. [CrossRef]

16. Veres, D.; Bazin, L.; Landais, A.; Toyé Mahamadou Kele, H.; Lemieux-Dudon, B.; Parrenin, F.; Martinerie, P.; Blayo, E.; Blunier, T.; Capron, E.; et al. The Antarctic ice core chornology (AICC2012): An optimized multi-parameter and multi-site dating approach for the last 120 thousands years. *Clim. Past* **2013**, *9*, 1733–1748. [CrossRef]

17. Baccolo, G.; Clemenza, M.; Delmonte, B.; Maffezzoli, N.; Nastasi, M.; Previtali, E.; Maggi, V. A new method based on Low Background Instrumental Neutron Activation Analysis for major, trace and ultra-trace elements determination in atmospheric mineral dust from polar ice cores. *Anal. Chim. Acta* **2016**, *922*, 11–18. [CrossRef] [PubMed]

18. Ruth, U.; Barbante, C.; Bigler, M.; Delmonte, B.; Fischer, H.; Gabrielli, P.; Gaspari, V.; Kaufmann, P.; Lambert, F.; Maggi, V.; et al. Proxies and measurement techinques for mineral dust in Antarctic ice cores. *Environ. Sci. Technol.* **2008**, *42*, 5675–5681. [CrossRef]

19. Quartieri, S. Synchrotron Radiation in the Earth Sciences. In *Synchrotron Radiation*; Springer: Berlin, Germany, 2014; pp. 641–660.

20. Hawkesworth, C.J.; Kemp, A.I.S. Evolution of the continental crust. *Nature* **2006**, *443*, 811–817. [CrossRef] [PubMed]

21. Marino, F.; Castellano, E.; Ceccato, D.; De Deckker, P.; Delmonte, B.; Gheramandi, G.; Maggi, V.; Petit, J.R.; Revel-Rolland, M.; Udisti, R. Defining the geochemical composition of the EPICA Dome C ice core dust during the last glacial-interglacial cycle. *Geochem. Geophys. Geosys.* **2008**, *9*. [CrossRef]

22. Boyd, P.W.; Ellwood, M.J. The biogeochemical cycle of iron in the ocean. *Nat. Geosci.* **2010**, *3*, 675–682. [CrossRef]

23. Wolff, E.W.; Fischer, H.; Fundel, F.; Ruth, U.; Twarloh, B.; Littot, G.C.; Mulvaney, R.; Rothlisberger, R.; de Angelis, M.; Boutron, C.F.; et al. Southern Ocean sea-ice extent, productivity and iron flux over the past eight glacial cycles. *Nature* **2006**, *440*, 491–496. [CrossRef] [PubMed]

24. Conway, T.M.; Wolff, E.W.; Röthlisberger, R.; Mulvaney, R.; Elderfield, H.E. Constraints on soluble aerosol iron flux to the Southern Ocean at the Last Glacial Maximum. *Nat. Commun.* **2015**, *6*. [CrossRef] [PubMed]

25. Baccolo, G.; Maffezzoli, N.; Clemenza, M.; Delmonte, B.; Prata, M.; Salvini, A.; Maggi, V.; Previtali, E. Low background neutron activation analysis: A powerful tool for atmopsheric mineral dust analyisis in ice cores. *J. Radioanal. Nucl. Chem.* **2015**, *306*. [CrossRef]

26. Macis, S.; Cibin, G.; Maggi, V.; Hampai, D.; Baccolo, G.; Delmonte, B.; D'Elia, A.; Marcelli, A. Microdrop deposition technique: Preparation and characterization of diluted suspended particulate samples. *Condens. Matter* **2018**, *3*, 21. [CrossRef]

27. Dent, A.J.; Cibin, G.; Ramos, S.; Smith, A.D.; Scott, S.M.; Varandas, L.; Pearson, M.R.; Krumpa, N.A.; Jones, C.P.; Robbins, P.E. B18: A core XAS spectroscopy beamline for Diamond. *J. Phys. Conf. Ser.* **2009**, *190*. [CrossRef]

28. Baccolo, G.; Delmonte, B.; Albani, S.; Baroni, C.; Cibin, G.; Frezzotti, M.; Hampai, D.; Marcelli, A.; Revel, M.; Salvatore, M.C.; et al. Regionalization of the atmospheric dust cycle on the periphery of the East Antarctic ice sheet since the last glacial maximum. *Geochem. Geophy. Geosy.* **2018**, in press.

29. Cibin, G.; Marcelli, A.; Maggi, V.; Sala, M.; Marino, F.; Delmonte, B.; Albani, S.; Pignotti, S. First combined total reflection X-ray fluorescence and grazing incidence X-ray absorption spectroscopy characterization of aeolian dust archived in Antarctica and Alpine deep ice cores. *Spectrochim. Acta B* **2008**, *63*, 1503–1510. [CrossRef]

30. Marcelli, A.; Hampai, D.; Giannone, F.; Sala, M.; Maggi, V.; Marino, F.; Pignotti, F.; Cibin, G. XRF-XANES characterization of deep ice core insoluble dust. *J. Anal. At. Spectrom.* **2012**, *27*, 33–37. [CrossRef]

31. Albani, S.; Delmonte, B.; Maggi, V.; Baroni, C.; Petit, J.R.; Stenni, B.; Mazzola, C.; Frezzotti, M. Interpreting last glacial to Holocene dust changes at Talos Dome (East Antarctica): Implications for atmospheric variations from regional to hemispheric scales. *Clim. Past* **2012**, *8*, 741–750. [CrossRef]

32. Delmonte, B.; Baroni, C.; Andersson, P.S.; Shoberg, H.; Hansson, M.; Aciego, S.; Petit, J.R.; Albani, S.; Mazzola, C.; Maggi, V.; et al. Aeolian dust in the Talos Dome ice core (East Antarctica, Pacific/Ross Sea sector): Victoria Land versus remote sources over the last two climatic cycle. *J. Quat. Sci.* **2010**, *25*, 1327–1337. [CrossRef]

33. Traversi, R.; Becagli, S.; Castellano, E.; Marino, F.; Rugi, F.; Severi, M.; de Angelis, M.; Fischer, H.; Hansson, M.; Stauffer, B.; et al. Sulfate spikes in the deep layers of EPICA-Dome C ice core: Evidence of glaciological artifacts. *Environ. Sci. Technol.* **2009**, *43*, 8737–8743. [CrossRef] [PubMed]

34. Delmonte, B.; Petit, J.R.; Maggi, V. Glacial to Holocene implications of the new 27000-year dust record from the EPICA Dome C (East Atarctica) ice core. *Clim. Dyn.* **2002**, *18*, 647–660.

35. Rudnick, R.L.; Gao, S. Composition of the continental crust. In *Treatise on Geochemistry*; Rudnick, R.L., Ed.; Elsevier: New York, NY, USA, 2003; Volume 3, pp. 1–64.

36. Legrand, M.; Mayewski, P. Glaciochemistry of polar ice cores: A review. *Rev. Geophys.* **1997**, *35*, 219–243. [CrossRef]

37. Rothlisberger, R.; Hutterli, A.; Sommer, S.; Wolff, E.W.; Mulvaney, R. Factors controlling nitrate in ice cores: Evidence from the Dome c deep ice core. *J. Geophys. Res.* **2000**, *105*, 20565–20572. [CrossRef]

38. Delmonte, B.; Andersson, P.S.; Shoberg, H.; Hansson, M.; Petit, J.R.; Delmas, R.; Gaiero, D.M.; Maggi, V.; Frezzotti, M. Geographic provenance of aeolian dust in East Antarctica during Pleistocene glaciations: Preliminary results from Talos Dome and comparison with East Antarctic and new Andean ice core data. *Quat. Sci. Rev.* **2010**, *29*, 256–264. [CrossRef]

39. Price, P.B. A habitat for psychrophiles in deep Antarctic ice. *PNAS* **2000**, *97*, 1247–1251. [CrossRef] [PubMed]

40. Berry, A.J.; O'Neill, H.S.C.; Jayasuriya, K.D.; Campbell, S.J.; Foran, G.J. XANES calibrations for the oxidation state of iron in a silicate glass. *Am. Mineral.* **2003**, *88*, 967–977. [CrossRef]

41. Chassé, M.; Griffin, W.L.; O'Reilly, S.Y.; Calas, G. Scandium speciation in a world-class lateritic deposit. *Geochem. Perspect. Lett.* **2017**, *3*, 105–114. [CrossRef]

42. Johnstone, S.G.; Burton, E.D.; Keene, A.F.; Planer-Friedrich, B.; Voegelin, A.; Blackford, M.G.; Lumpkin, G.R. Arsenic mobilization and iron transformations during sulfidization of As (V)-bearing jarosite. *Chem. Geol.* **2012**, *334*, 9–24. [CrossRef]

43. Wilke, M.; Farges, F.; Petit, P.E.; Brown, G.E., Jr.; Martin, F. Oxidation state and coordination of Fe in minerals: An Fe K-XANES spectroscopic study. *Am. Mineral.* **2001**, *86*, 714–730. [CrossRef]

44. Giuli, G.; Eeckhout, S.G.; Paris, E.; Koeberl, C.; Pratesi, G. Iron oxidation state in impact glass from the K/T boundary at Beloc, Haiti, by high-resolution XANES spectroscopy. *Meteorit. Planet. Sci.* **2005**, *40*, 1575–1580. [CrossRef]

45. Formenti, P.; Caquineau, S.; Chevaillier, S.; Klaver, A.; Desboeufs, K.; Rajot, J.L.; Belin, S.; Briois, V. Dominance of goethite over hematite in iron oxides of mineral dust from Western Africa: Quantitative partitioning by X-ray absorption spectroscopy. *J. Geophys. Res. Atmos.* **2014**, *119*, 12740–12754. [CrossRef]

46. Anesio, A.M.; Laybourn-Parry, J. Glaciers and ice sheets as a biome. *Trends Ecol. Evol.* **2012**, *27*, 219–225. [CrossRef] [PubMed]

47. Elwood Madden, M.E.; Bodnar, R.J.; Rimstidt, J.D. Jarosite as an indicator of water-limited chemical weathering on Mars. *Nature* **2004**, *431*, 821–823. [CrossRef] [PubMed]

48. Spolaor, A.; Vallelonga, P.; Cozzi, G.; Gabrieli, J.; Varin, C.; Kehrwald, N.; Zennaro, P.; Boutron, C.; Barbante, C. Iron speciation in aerosol dust influences iron bioavailability over glacial-interglacial timescales. *Geophys. Res. Lett.* **2013**, *40*, 1618–1623. [CrossRef]

Article

The Study of Characteristic Environmental Sites Affected by Diverse Sources of Mineral Matter Using Compositional Data Analysis

Antonio Speranza *, Rosa Caggiano, Giulia Pavese and Vito Summa

IMAA, Istituto di Metodologie per l'Analisi Ambientale, CNR, 85050 Tito Scalo, PZ, Italy;
rosa.caggiano@imaa.cnr.it (R.C.); giulia.pavese@imaa.cnr.it (G.P.); vito.summa@imaa.cnr.it (V.S.)
* Correspondence: antonio.speranza@imaa.cnr.it; Tel.: +39-0971-427-230

Received: 13 March 2018; Accepted: 3 May 2018; Published: 7 May 2018

Abstract: Compositional data analysis was applied on mineral element concentrations (i.e., Al, Ti, Si, Ca, Mg, Fe, Sr) content in PM_{10}, $PM_{2.5}$ and PM_1 simultaneous measurements at three characteristic environmental sites: kerbside, background and rural site. Different possible sources of mineral trace elements affecting the PM in the considered sites were highlighted. Particularly, results show that compositional data analysis allows for the assessment of chemical/physical differences between mineral element concentrations of PM. These differences can be associated with both different kinds of involved mineral sources and different mechanisms of accumulation/dispersion of PM at the considered sites.

Keywords: particulate matter; simultaneous measurements; mineral elements; compositional data analysis

---

1. Introduction

Particulate matter (PM) is a mixture of particles present in the air with different masses, sizes, shapes, surface areas, characterized by many chemical elements and compounds. The interest in PM has widely increased in the last several decades owing to its detrimental effect on the public health and the environment [1–3]. PM particles when inhaled can be deposited within the respiratory system and can act as a universal carrier of a wide variety of chemical substances potentially toxic [4,5]. It is well known that acute and chronic exposure to PM is associated with adverse effects on the cardiovascular and respiratory system [6]. PM is also associated with air quality and visibility degradation as well as to Earth's climate change [7,8]. PM particles can affect solar radiation as it passes through the atmosphere due to their properties of scattering and absorptions of light as well as being able to alter the radiative properties of weather clouds and their lifetime determining a possible effect on incoming and outgoing solar radiation and on the overall energy balance of Earth [9–11].

In the light of aforesaid significance of PM, effective mitigation strategies of PM levels into the air, to protect the environment and the public health, require a detailed assessment of the possible PM emission sources in relation to its chemical composition [12]. In the European context, selected groups of chemical elements and compounds have been linked to specific natural and anthropogenic sources of PM such as Al, Si, Ca, Fe, Ti, Mg and Sr to crustal and mineral matter and African dust, Na, Cl and Mg to marine sources and sea spray, V and Ni to fuel-oil combustion, SO_4^{2-}, NO_3^- and NH_4^+ to secondary aerosol and long range transport [13]. However, the identification of an element or a group of elements that can unequivocally be attributed to specific natural or anthropogenic sources of mineral matter has proven to be problematic. Sources of mineral matter including desert, crustal dust due to road traffic and farming activities as well as dust due to demolition and construction activities can have in common the same range of chemical elements. Up to today, the identification and characterization of different sources of mineral matter contributing to atmospheric PM mixture remains inadequate and requires still further research as pointed out in the literature [13–15].

In the meanwhile, several authors have focused their studies on the simultaneous measurements of PM size fractions: PM_{10}, $PM_{2.5}$ and PM_1 (i.e., particles with aerodynamic diameters below 10, 2.5 and 1 μm, respectively). Due to the fact that PM particles with different sizes are emitted into the air from different sources and have different physical and chemical characteristics, the assessment of the PM size fractions can provide with important data on the process of formation of the PM and the identification of its sources. The PM coarse size fraction (i.e., particles with aerodynamic diameters between 2.5 and 10 μm) can mainly originate from both natural and anthropogenic sources including desert dust, volcanic eruptions, sea spray, fugitive dust from paved and unpaved dusty roads, demolition and construction activities [16–18]. The fine size fraction ($PM_{2.5}$) and submicron size fraction (PM_1) can mainly originate from anthropogenic sources such as industrial activities, road traffic, different kinds of combustion processes and secondary particles generated in the atmosphere [19–21]. The simultaneous measurements of PM_{10}, $PM_{2.5}$ and PM_1 have been carried out in a range of different environmental sites such as traffic point [22], urban traffic and suburban background sites [23], Nordic background site and wild fire episodes [24], urban background and rural sites [25,26], industrial sites [27,28], superstation site [29], urban areas and surroundings sites [30] and suburban sites and regional background [31]. Further environmental investigations on size-segregate PM simultaneous measurements were reported elsewhere [32,33].

This preliminary study investigates the application of compositional data analysis to mineral element concentrations of size-segregated PM simultaneous measurements. Compositional data consist of vectors whose components represent proportions or percentages of a certain quantity. The characteristic of these vectors is that the sum of their components is a constant, equal to 1 for proportions, 100 for percentages or some other constant c. These compositional data may refer to mineral composition of rocks, sediments, pollutant compositions, mixture of gases, water composition, etc. Compositional data with appropriate statistical tools have been used for interpreting environmental data as well as for characterizing processes acting in the environment. Statistical analysis of compositional data began with Aitchison [34,35] and since then has undergone several developments. Today, a consolidated statistical technique is considered [36–40].

This study is based on mineral element concentrations of PM_{10} $PM_{2.5}$ and PM_1 simultaneous measurements sampled in three characteristic sites (i.e., kerbside, background and rural site) possibly affected by diverse sources of mineral matter [41]. The relevance of mineral element concentrations of PM at kerbside site was taken into consideration. Mineral element concentrations of PM from the kerbside were compared with that from background and rural sites [42,43].

The main objective of this preliminary study is to evaluate the essential differences between the two patterns of variability of kerbside and background site dataset as well as between the two patterns of variability of kerbside and rural site dataset. These comparisons aim to highlight the underlying processes that influence the mineral element concentrations of PM of the considered environmental sites. The present study shows that the PM of the investigated environmental sites can be affected from different sources of mineral matter. Moreover, mechanisms of accumulation and/or dispersion of mineral matter can be observed.

2. Materials and Methods

Mineral elements concentrations of PM_{10}, $PM_{2.5}$ and PM_1 simultaneous measurements as reported in literature have been considered and they refer to three characteristics monitored environmental sites namely rural, background and kerbside site. The kerbside site was placed close to a trafficked road and at the side of a street canyon, the background site was placed in a residential area close to school building and the rural site was located away from direct emission sources and in a site surrounded by fields. The measurements were conducted in winter (January–February). The samples of PM were collected using a rotating drum impactor. The elements on the collected samples were analyzed using a synchrotron radiation-induced X-ray fluorescence spectrometry [41]. The mineral elements considered were Al, Ti, Si, Ca, Mg, Fe, Sr. These mineral elements were mostly and commonly interpreted as related to mineral matter [13]. In this section, the methods used for the compositional data analysis are summarized.

2.1. Compositional Data and Sample Space

Compositional data are vectors whose components are positive numbers and they sum to a constant, usually 1 or 100. The sample space of a compositional observation **x** with three components is the unit simplex:

$$S_c^3 = \left\{ \mathbf{x} = (x_1, x_2, x_3) \,\middle|\, x_j > 0, j = 1, 2, 3; x_1 + x_2 + x_3 = c \right\}. \tag{1}$$

The simultaneous measurements PM_{10}, $PM_{2.5}$ and PM_1 can be decomposed in their relative fractions as coarse (see Equation (2)), intermodal (see Equation (3)) and submicron, PM_1, mass concentration [44–46]:

$$PM_{10-2.5} = PM_{10} - PM_{2.5}, \tag{2}$$

$$PM_{2.5-1} = PM_{2.5} - PM_1. \tag{3}$$

These fractions can be converted into compositions based on proportions by weight [40,47]:

$$\mathbf{x} = \left(\frac{PM_{10-2.5}}{PM_{10}}, \frac{PM_{2.5-1}}{PM_{10}}, \frac{PM_1}{PM_{10}} \right) \%. \tag{4}$$

The compositional variables of this vector, **x**, are nonnegative and they sum to a constant $c = 100$ (see Equation (1)). The compositional data related to simultaneous sampling of PM_{10}, $PM_{2.5}$ and PM_1 of various mineral elements can be cast into the form of a matrix **x**, with i rows representing the mineral elements and j columns representing the coarse, intermodal and submicron size fractions. This matrix, a three part compositional dataset, takes into account the mineral element concentrations of PM with respect to its coarse, intermodal, and submicron, size fraction in %.

The following matrix for the simultaneous measurements PM_{10}, $PM_{2.5}$ and PM_1 of the elements Al, Ti, Si, Ca, Mg, Fe, Sr was considered in each characteristic environmental sites (i.e., rural, kerbside and background):

$$\mathbf{x} = \left\{ \begin{array}{ccc} \left(\dfrac{PM_{10-2.5}}{PM_{10}}\right)_{Al} & \left(\dfrac{PM_{2.5-1}}{PM_{10}}\right)_{Al} & \left(\dfrac{PM_1}{PM_{10}}\right)_{Al} \\[2mm] \left(\dfrac{PM_{10-2.5}}{PM_{10}}\right)_{Ti} & \left(\dfrac{PM_{2.5-1}}{PM_{10}}\right)_{Ti} & \left(\dfrac{PM_1}{PM_{10}}\right)_{Ti} \\[2mm] \left(\dfrac{PM_{10-2.5}}{PM_{10}}\right)_{Si} & \left(\dfrac{PM_{2.5-1}}{PM_{10}}\right)_{Si} & \left(\dfrac{PM_1}{PM_{10}}\right)_{Si} \\[2mm] \left(\dfrac{PM_{10-2.5}}{PM_{10}}\right)_{Ca} & \left(\dfrac{PM_{2.5-1}}{PM_{10}}\right)_{Ca} & \left(\dfrac{PM_1}{PM_{10}}\right)_{Ca} \\[2mm] \left(\dfrac{PM_{10-2.5}}{PM_{10}}\right)_{Mg} & \left(\dfrac{PM_{2.5-1}}{PM_{10}}\right)_{Mg} & \left(\dfrac{PM_1}{PM_{10}}\right)_{Mg} \\[2mm] \left(\dfrac{PM_{10-2.5}}{PM_{10}}\right)_{Fe} & \left(\dfrac{PM_{2.5-1}}{PM_{10}}\right)_{Fe} & \left(\dfrac{PM_1}{PM_{10}}\right)_{Fe} \\[2mm] \left(\dfrac{PM_{10-2.5}}{PM_{10}}\right)_{Sr} & \left(\dfrac{PM_{2.5-1}}{PM_{10}}\right)_{Sr} & \left(\dfrac{PM_1}{PM_{10}}\right)_{Sr} \end{array} \right\} \%. \tag{5}$$

2.2. Transformation of Compositional Data

Compositional data in Equation (4) are constrained to a constant sum. The statistical analysis of compositional data requires an approach based on log-ratios transformation. Using this transformation, a composition is represented as a real vector. The compositional data are transformed into coordinates using *ilr* (isometric log-ratio) transformations [48,49] (see Equation (6)):

$$ilr(\mathbf{x}) = \left(\frac{1}{\sqrt{2}} \ln\left(\frac{PM_{10-2.5}}{PM_{2.5-1}} \right), \frac{1}{\sqrt{6}} \ln\left(\frac{PM_{10-2.5}\,PM_{2.5-1}}{PM_1^2} \right) \right) = (ilr_1, ilr_2) = \mathbf{y}. \tag{6}$$

The following matrix for PM_{10}, $PM_{2.5}$ and PM_1 simultaneous measurements of the elements Al, Ti, Si, Ca, Mg, Fe, Sr was considered in each characteristic environmental sites (i.e., rural, kerbside and background) in terms of log-ratios:

$$Y = \left\{ \begin{array}{cc} ilr_{1\,Al} & ilr_{2\,Al} \\ ilr_{1\,Ti} & ilr_{2\,Ti} \\ ilr_{1\,Si} & ilr_{2\,Si} \\ ilr_{1\,Ca} & ilr_{2\,Ca} \\ ilr_{1\,Mg} & ilr_{2\,Mg} \\ ilr_{1\,Fe} & ilr_{2\,Fe} \\ ilr_{1\,Sr} & ilr_{2\,Sr} \end{array} \right\}. \tag{7}$$

The isometric coordinates ilr_1 and ilr_2 can be inverse transformed by:

$$\mathbf{x} = C\left(\exp\left(\tfrac{ilr_2}{\sqrt{6}} + \tfrac{ilr_1}{\sqrt{2}} \right), \ \exp\left(\tfrac{ilr_2}{\sqrt{6}} - \tfrac{ilr_1}{\sqrt{2}} \right), \ \exp\left(-\tfrac{2ilr_2}{\sqrt{6}} \right) \right), \tag{8}$$

where C is the closure operation for a vector $\mathbf{x}$ defined as below in Equation (9). This operation divides each component of the vector $\mathbf{x}$ by the sum of its components, hence scaling the vector to the constant c:

$$C(\mathbf{x}) = \left(\frac{cx_1}{x_1 + x_2 + x_3}, \frac{cx_2}{x_1 + x_2 + x_3}, \frac{cx_3}{x_1 + x_2 + x_3} \right). \tag{9}$$

2.3. Triangular Diagram Representation, Centering and Rescaling Technique

The compositional datasets, their centres and confidence regions can be represented using a triangular diagram. The data is displayed by Graham and Midgley [50]. Calculations were produced using Coda Pack Software [51] and R Software [52]. Compositional data can be centered using the perturbation operator of the simplex [35].

The perturbation operation is defined as the perturbation $\mathbf{p}$ applied to a composition $\mathbf{x}$ that produces the composition $\mathbf{v}$:

$$\mathbf{v} = \mathbf{p} \oplus \mathbf{x} = (p_1, p_2, p_3) \oplus (x_1, x_2, x_3) = C(p_1 x_1, p_2 x_2, p_3 x_3), \tag{10}$$

with $\mathbf{v}$, $\mathbf{p}$ and $\mathbf{x}$ vectors in S^3_c; C is the closure operation (see Equation (9)).

Perturbing a vector $\mathbf{x}$ by its inverse (see Equation (11)), it is possible to locate any composition in the baricenter of the triangular diagram:

$$\mathbf{x}^{-1} = \left(x_1^{-1}, \ x_2^{-1}, \ x_3^{-1} \right). \tag{11}$$

Likewise, it is possible to centre a compositional dataset of size n:

$$\left\{ \left(x_{i1}, \ x_{i2} \ \ x_{i3} \right), i = 1, 2, \ldots, n \right\}, \tag{12}$$

using the inverse of its centre $\mathbf{g}^{-1}$ defined as Equation (13) [53,54]:

$$\mathbf{g} = C\left(\ g_1,\ \ g_2,\ \ g_3\ \right)\ \text{where}\ g_j = \left(\prod_{i=1}^{n} x_{i,j}\right)^{\frac{1}{n}}, j = 1, 2, 3. \tag{13}$$

The centering and rescaling of the compositional data allows for better visualizing compositions close to the boundary of the triangular diagram preserving straight lines of the grid as well as statistical properties [55].

2.4. Perturbation Difference

The perturbation difference is defined as the perturbation $\mathbf{p}$ to which a change can be attributed from a composition $\mathbf{x}$ to a composition $\mathbf{y}$, whatever the processes involved, (see Equation (14))

$$\mathbf{p} = \mathbf{xz} = (x_1, x_2, x_3)(z_1, z_2, z_3) = C\left(\frac{x_1}{z_1}, \frac{x_2}{z_2}, \frac{x_3}{z_3}\right), \tag{14}$$

with $\mathbf{p}$, $\mathbf{x}$ and $\mathbf{z}$ vectors in S^3_c [56]; C is the closure operation (see Equation (9)).

2.5. Testing Hypothesis of Multivariare Normal Distribution

The multivariate normal distribution is a generalization of the normal distribution to higher dimensions. The test hypothesis on multivariate normal distribution of the compositional dataset relating to rural, kerbside and background (see Equation (5)) is needed before performing the test hypothesis about center and covariance structure. The test used for multivariate normality is the Anderson–Darling, Cramer–von Mises and Watson of log-ratios transformed dataset (see Equation (7)) [35,36].

2.6. Testing Hypothesis about the Center and the Covariance Structure

In order to evaluate whether there are real differences between two datasets in the center, in the covariance structure, or in both, a test hypothesis about center and covariance structure was performed. The test is applied to log-ratios transformed dataset (see Equation (7)). The methods are described in [35] (pp. 153–158), [36]. The PM size fractions and its chemical composition provide with important data on the process of formation of the PM and the identification of its sources [16–21]. The center and the covariance structure of a compositional dataset $\mathbf{x}$ as derived from PM simultaneous measurements are linked to the chemical composition and distribution of the elements within coarse, intermodal and submicron size fraction. The test hypothesis about center and covariance structure allows for the statistical evaluation of differences between the chemical composition and the size distribution between two datasets. Therefore, the results of the above test provide information about the diverse PM sources and its formation processes between two considered datasets.

3. Results and Discussion

The difference between the mineral element concentrations of PM relating to background and kerbside site was evaluated comparing their compositional datasets. Figure 1a shows the three part compositional dataset of two characteristic monitored environmental sites such as background and kerbside after the application of the centering and rescaling technique. The three part compositional dataset of background and kerbside are displayed with high values of coarse size fraction above 55% and with low values of intermodal and submicron size fraction, which are below about 35% and 12%, respectively. The coarse size fraction is the dominant component in mineral tracers [57]. The application of the centering and rescaling technique allows for a better visualization of the data despite the low values of intermodal and submicron size fraction. The three compositional datasets and their respective centres are closely displayed. Thus, in order to confirm or reject the hypothesis about the occurrence of two distinct sets, a statistical analysis is performed [57].

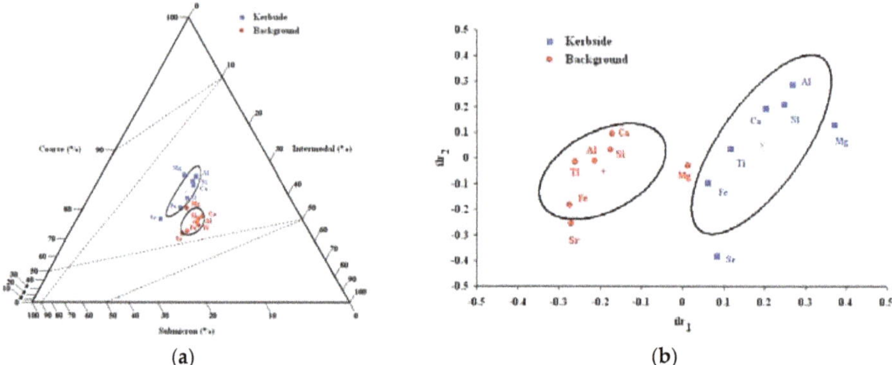

(a)   (b)

Figure 1. The mineral elements Al, Ti, Si, Ca, Mg, Fe, Sr; (**a**) distribution of the compositional data plotted using Equation (4) and after perturbation $\mathbf{g}^{-1}$ = (8.53, 20.84, 70.63). The centre for background site is at $\mathbf{g}$ = (59.31, 31.93, 8.76) denoted with +, and the centre for kerbside site is at $\mathbf{g}$ = (70.91, 22.08, 7.01) denoted with ×; (**b**) distribution of the data as for isometric log-ratios (see Equation (6)) after perturbation. The centre for both datasets is at (ilr_1,ilr_2) = (0,0), the centre for background site is at (ilr_1,ilr_2) = (−0.19, −0.05) denoted with +, and the centre for kerbside site is at (ilr_1,ilr_2) = (0.19,0.05) denoted with ×. The continuous lines are the confidence regions $(1 - \alpha)100\%$, $\alpha = 0.05$.

The two datasets of background and kerbside are tested for multivariate normality. The numerical results are reported in Table 1 and are compared with critical values reported by Stephens [36,58]. The bivariate angle test shows that the hypothesis of normality can be accepted for both background and kerbside site with a significance level greater than 10%.

Table 1. Tests on multivariate normality for the datasets of rural, kerbside and background sites.

Sites	Anderson–Darling	Cramer–von Mises	Watson
Rural			
ilr_1 marginal distribution	0.6727	0.1260	0.1244
ilr_2 marginal distribution	0.8657	0.1405	0.1377
Bivariate angle test statistics	1.7077	0.2927	0.0608
Kerbside			
ilr_1 marginal distribution	0.2821	0.0420	0.0419
ilr_2 marginal distribution	0.4277	0.0646	0.0633
Bivariate angle test statistics	1.6273	0.3314	0.1293
Background			
ilr_1 marginal distribution	0.6848	0.1024	0.1004
ilr_2 marginal distribution	0.4985	0.0951	0.0944
Bivariate angle test statistics	0.7345	0.0804	0.0510

The marginal test shows that the hypothesis of normality can be accepted for the kerbside site, referring to both ilr_1 and ilr_2, with a significance level greater than 10%. The marginal test shows that the hypothesis of normality can be accepted for the background site with a significance level greater than 5% and 10% for ilr_1 and ilr_2, respectively. Therefore, for each dataset, the hypothesis of multivariate normality cannot be rejected. Thus, the datasets of kerbside and background site are tested for hypothesis of equality in their centres and covariance structures [35] (p. 153), [36]. The results are reported in Table 2. The test value for the datasets related to kerbside and background site is below the critical value for the considered hypothesis of inequality of centres, equivalent to $\mu_1 \neq \mu_2$, and equality of covariance structure, equivalent to $\Sigma_1 = \Sigma_2$, thus this last hypothesis cannot be rejected. The inequality between the two centres indicates that the mineral element distribution differs for the two datasets. Mineral elements are more abundant in the coarse and intermodal size fraction in the dataset related to kerbside. Thus, it can be assumed that at the kerbside site there are mechanisms that promote the accumulation of elements in the coarser fractions.

Table 2. Test about the centres and the covariance structures for kerbside and background site.

Hypothesis	Test Value	χ^2 Critical Value ($\alpha = 0.05$)	Degrees of Freedom	Significance
$\mu_1 = \mu_2, \Sigma_1 = \Sigma_2$	25.57	11.07	5	0.0001
$\mu_1 \neq \mu_2, \Sigma_1 = \Sigma_2$	2.70	7.81	3	0.4403
$\mu_1 = \mu_2, \Sigma_1 \neq \Sigma_2$	-	5.99	2	-

However, the equality between the two covariance structures suggests that the datasets related to kerbside and background site cannot be regarded as clearly distinct for chemical composition (see Figure 2b).

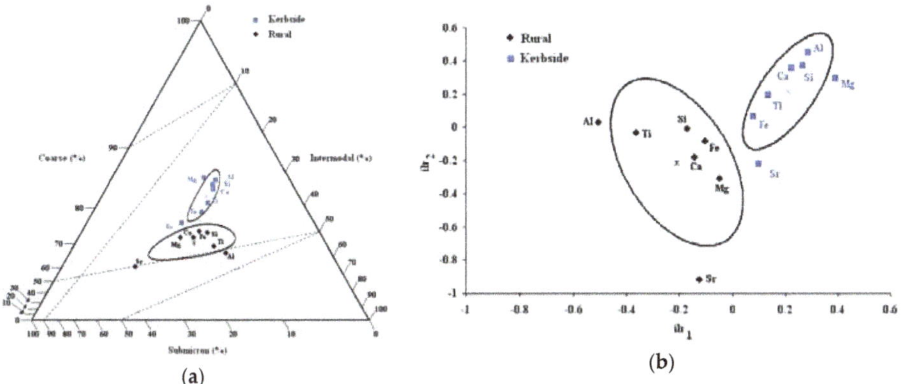

Figure 2. The mineral elements Al, Ti, Si, Ca, Mg, Fe, Sr; **(a)** distribution of the compositional data plotted using Equation (4) and after perturbation after perturbation g^{-1} = (9.9, 23.63, 66.47). The centre for the rural site is at g = (55.94, 31.52, 12.54) denoted with *, and the centre for kerbside site is at g = (70.91, 22.08, 7.01) denoted with ×; **(b)** distribution of the data as for isometric log-ratios (see Equation (6)) after perturbation. The centre for both dataset is at (ilr_1, ilr_2) = (0,0), the centre for the rural site is at (ilr_1, ilr_2) = $(-0.21, -0.21)$ denoted with *, and the centre for kerbside site is at (ilr_1, ilr_2) = (0.21,0.21) denoted with ×. The continuous lines are the confidence regions $(1 - \alpha)100\%$, α = 0.05.

These results can be interpreted so that kerbside and background site are characterized by similar sources of mineral matter for the set of considered mineral elements.

In order to evaluate the nature of the difference between the element concentrations of PM for kerbside and background site, the perturbation difference is calculated between the perturbation centres related to kerbside and background site compositional dataset [59]. The perturbation centre for background site is $(59.31, 31.93, 8.77)_{(background)}$. The perturbation centre for kerbside site is $(70.91, 22.08, 7.01)_{(kerbside)}$. The perturbation difference is $(44.51, 25.75, 29.74)_{(kerbside)-(background)}$ suggesting that the relative difference between these two sites is mainly in the coarse size fraction.

The simultaneous PM measurements at a kerbside site refer to a very busy road with a street canyon feature. Thus, it can be supposed that, at the kerbside, the effect of large traffic volumes combined with possibly narrow streets and multi-floor institutional and commercial building can determine a local characteristic accumulation/dispersion condition (e.g., street canyon environment) [60]. These conditions may lead to higher tracer concentrations of the coarse size fraction.

Likewise, the difference between the mineral elements concentrations of PM relating to kerbside and rural site was evaluated comparing their compositional datasets. Figure 2a shows the three part compositional dataset of two characteristic monitored environmental sites such as kerbside and rural site after the application of the centering and rescaling technique. The three part compositional data of kerbside and rural site are displayed with high values of coarse size fraction above approximately 50% and with low values of intermodal and submicron size fraction, which are below about 40% and 25%, respectively. The coarse size fraction is the dominant component. The application of the centering and rescaling technique allows for a better visualization of the data despite the low values of intermodal and submicron size fraction. The compositional datasets of kerbside and rural as well as their respective centres are closely displayed. As above, in order to confirm or reject the hypothesis about the occurrence of two distinct sets, a statistical analysis is performed. The dataset of rural site is tested for multivariate normality. The numerical results are reported in Table 1 and they are compared with critical values. The bivariate angle test shows that the hypothesis of normality can be accepted with a significance level greater than 10%. The marginal test shows that the hypothesis of normality can be accepted for the rural site with a significance level greater than 5% and 2.5% for ilr_1 and ilr_2, respectively. Therefore, the hypothesis of multivariate normality cannot be rejected. Thus, the datasets of kerbside and rural site are tested for hypothesis of equality in their centres and covariance structures. The results are reported in Table 3. The test values for the datasets related to kerbside and rural site are above the critical ones for each considered hypothesis, thus the equality of centres, equivalent to $\mu_1 = \mu_2$, of covariance structures, equivalent to $\Sigma_1 = \Sigma_2$, or of both has to be rejected. The two datasets related to kerbside and rural site have to be regarded as clearly distinct for chemical composition and mineral element distributions (see Figure 2b).

Table 3. Test about the centres and the covariance structures for kerbside and rural site.

Hypothesis	Test Value	χ^2 Critical Value ($\alpha = 0.05$)	Degrees of Freedom	Significance
$\mu_1 = \mu_2$, $\Sigma_1 = \Sigma_2$	27.26	11.07	5	0
$\mu_1 \neq \mu_2$, $\Sigma_1 = \Sigma_2$	8.42	7.81	3	0.0381
$\mu_1 = \mu_2$, $\Sigma_1 \neq \Sigma_2$	17.72	5.99	2	0.0001

These results can be interpreted as follows. The sources of mineral element at the rural site and at the kerbside site may differ. Moreover, the mechanism of accumulation and dispersion at the two sites may be also different (e.g., long range transport, dust resuspension and traffic-related processes). The combination of possibly different sources of mineral elements and diverse mechanism of accumulation/dispersion can concur to determine a different chemical composition and size distribution of the mineral matter contained in the PM.

In order to evaluate the nature of the difference between the mineral element concentrations of PM of kerbside and rural site, the perturbation difference is calculated between the perturbation centres related kerbside and rural site compositional dataset. The perturbation centre for kerbside site is $(70.91, 22.08, 7)_{(kerbside)}$, whereas the perturbation centre for rural site is $(55.94, 31.51, 12.55)_{(rural)}$. The perturbation difference is $(50.17, 27.74, 22.09)_{(kerbside)-(rural)}$ suggesting that the relative difference between these two sites is mainly in the coarse size fraction of the considered set of mineral elements. This may be a result of the resuspended mineral matter due to road traffic, which contributes to higher concentrations in the coarse size fraction at the kerbside site [61].

4. Conclusions

The statistical methods used for the analysis of compositional data allow for statistically validating either differences or similarities between the investigated datasets of the related environmental sites. These differences or similarities can be associated to both the kind of involved mineral sources and possible mechanisms of addition and/or subtraction of materials that influences the behavior of the characteristic environmental sites.

Results highlight that the datasets of kerbside site and background site have different centers and equal covariance structures. The mineral elements of PM of these two sites are compositionally equivalent. Though the two distinct centres indicate that mineral elements have different distribution. This can be related to possible different mechanisms of accumulation and/or dispersion of PM at the two sites. At the kerbside, the traffic in combination narrow streets and multi-floor building determines local characteristic conditions, which lead to higher tracer concentrations of the coarse size fraction.

The datasets of kerbside site and the rural site have different centers and covariance structures. The mineral elements of PM of these two sites are different for composition and size distribution. These two sites have different sources of mineral elements. Furthermore, the mechanism of accumulation and/or dispersion at the two sites may also be different. The combination of different sources and diverse mechanisms of accumulation and/or dispersion can concur to determine a different chemical composition and size distribution of the mineral elements of PM.

The compositional analysis applied to mineral element concentrations of PM_{10}, $PM_{2.5}$ and PM_1 simultaneous measurements is a technique that can be used to study environmental sites interested by different sources of mineral matter.

Author Contributions: A.S. provided the idea and designed the study. A.S., R.C., G.P. and V.S. illustrated the figures and wrote the manuscript.

Conflicts of Interest: The authors declare no conflict of interest.

References

1. Anderson, J.O.; Thundiyil, J.G.; Stolbach, A. Clearing the air: A review of the effects of particulate matter air pollution on human health. *J. Med. Toxicol.* **2012**, *8*, 166–175. [CrossRef] [PubMed]

2. Janssen, N.A.H.; World Health Organization. *Health Effects of Black Carbon*; World Health Organization, Regional Office for Europe: Copenhagen, Denmark, 2012; ISBN 978-92-890-0265-3.

3. WHO (World Health Organization). *Air Quality Guidelines: Global Update 2005: Particulate Matter, Ozone, Nitrogen Dioxide, and Sulfur Dioxide*; World Health Organization, Regional Office for Europe: Geneva, Switzerland, 2006.

4. Kelly, F.J.; Fussell, J.C. Size, source and chemical composition as determinants of toxicity attributable to ambient particulate matter. *Atmos. Environ.* **2012**, *60*, 504–526. [CrossRef]

5. Heyder, J. Deposition of Inhaled Particles in the Human Respiratory Tract and Consequences for Regional Targeting in Respiratory Drug Delivery. *Proc. Am. Thorac. Soc.* **2004**, *1*, 315–320. [CrossRef] [PubMed]

6. Pope, C.A., III; Dockery, D.W. Health effects of fine particulate air pollution: Lines that connect. *J. Air Waste Manag. Assoc.* **2006**, *56*, 709–742. [CrossRef] [PubMed]

7. Caggiano, R.; D'Emilio, M.; Macchiato, M.; Ragosta, M. Experimental and statistical investigations on atmospheric heavy metals concentrations in an industrial area of Southern Italy. *Nuovo Cimento C* **2001**, *24*, 391–406.

8. Stocker, T.F.; Qin, D.; Plattner, G.K.; Tignor, M.M.; Allen, S.K.; Boschung, J.; Nauels, A.; Xia, Y.; Bex, V.; Midgley, P.M. *Climate Change 2013: The Physical Science Basis. Contribution of Working Group I to the Fifth Assessment Report of IPCC the Intergovernmental Panel on Climate Change*; Cambridge University Press: Cambridge, UK, 2014; ISBN 978-1-107-66182-0.

9. Prospero, J.M.; Charlson, R.J.; Mohnen, V.; Jaenicke, R.; Delany, A.C.; Moyers, J.; Rahn, K. The atmospheric aerosol system: An overview. *Rev. Geophys.* **1983**, *21*, 1607–1629. [CrossRef]

10. Haywood, J.; Boucher, O. Estimates of the direct and indirect radiative forcing due to tropospheric aerosols: A review. *Rev. Geophys.* **2000**, *38*, 513–543. [CrossRef]

11. Prospero, J.M. African dust: Its large-scale transport over the Atlantic Ocean and its impact on the Mediterranean region. In *Regional Climate Variability and Its Impacts in The Mediterranean Area*; Springer: Dordrecht, The Netherlands, 2007; pp. 15–38. ISBN 978-1-4020-6429-6.

12. Putaud, J.P.; Van Dingenen, R.; Alastuey, A.; Bauer, H.; Birmili, W.; Cyrys, J.; Flentje, H.; Fuzzi, S.; Gehrig, R.; Hansson, H.C.; et al. A European aerosol phenomenology–3: Physical and chemical characteristics of particulate matter from 60 rural, urban, and kerbside sites across Europe. *Atmos. Environ.* **2010**, *44*, 1308–1320. [CrossRef]

13. Viana, M.; Kuhlbusch, T.A.J.; Querol, X.; Alastuey, A.; Harrison, R.M.; Hopke, P.K.; Winiwarter, W.; Vallius, M.; Szidat, S.; Prévôt, A.S.H.; et al. Source apportionment of particulate matter in Europe: A review of methods and results. *J. Aerosol Sci.* **2008**, *39*, 827–849. [CrossRef]

14. Thorpe, A.; Harrison, R.M. Sources and properties of non-exhaust particulate matter from road traffic: A review. *Sci. Total Environ.* **2008**, *400*, 270–282. [CrossRef] [PubMed]

15. Pant, P.; Harrison, R.M. Estimation of the contribution of road traffic emissions to particulate matter concentrations from field measurements: A review. *Atmos. Environ.* **2013**, *77*, 78–97. [CrossRef]

16. Van Dingenen, R.; Raes, F.; Putaud, J.P.; Baltensperger, U.; Charron, A.; Facchini, M.C.; Decesari, S.; Fuzzi, S.; Gehrig, R.; Hansson, H.C.; et al. European aerosol phenomenology-1: Physical characteristics of particulate matter at kerbside, urban, rural and background sites in Europe. *Atmos. Environ.* **2004**, *38*, 2561–2577. [CrossRef]

17. Colbeck, I. *Environmental Chemistry of Aerosols*; Blackwell Publishing Ltd.: Oxford, UK, 2008; ISBN 9781405139199.

18. Liu, Y.; Zhang, S.; Fan, Q.; Wu, D.; Chan, P.; Wang, X.; Fan, S.; Feng, Y.; Hong, Y. Accessing the Impact of Sea-Salt Emissions on Aerosol Chemical Formation and Deposition over Pearl River Delta, China. *Aerosol Air Qual. Res.* **2015**, *15*, 2232–2245. [CrossRef]

19. Caggiano, R.; Macchiato, M.; Trippetta, S. Levels, chemical composition and sources of fine aerosol particles (PM$_1$) in an area of the Mediterranean basin. *Sci. Total Environ.* **2010**, *408*, 884–895. [CrossRef] [PubMed]

20. Chakraborty, A.; Gupta, T. Chemical characterization and source apportionment of submicron (PM1) aerosol in Kanpur region, India. *Aerosol Air Qual. Res.* **2010**, *10*, 433–445. [CrossRef]

21. Margiotta, S.; Lettino, A.; Speranza, A.; Summa, V. PM 1 geochemical and mineralogical characterization using SEM-EDX to identify particle origin–Agri Valley pilot area (Basilicata, southern Italy). *Nat. Hazards Earth Syst. Sci.* **2015**, *15*, 1551–1561. [CrossRef]

22. Rogula-Kozłowska, W.; Rogula-Kupiec, P.; Mathews, B.; Klejnowski, K. Effects of road traffic on the ambient concentrations of three PM fractions and their main components in a large Upper Silesian city. *Ann. Wars. Univ. Life Sci.-SGGW Land Reclam.* **2013**, *45*, 243–253. [CrossRef]

23. Matassoni, L.; Pratesi, G.; Centioli, D.; Cadoni, F.; Lucarelli, F.; Nava, S.; Malesani, P. Saharan dust contribution to PM$_{10}$, PM$_{2.5}$ and PM$_1$ in urban and suburban areas of Rome: A comparison between single-particle SEM-EDS analysis and whole-sample PIXE analysis. *J. Environ. Monit.* **2011**, *13*, 732–742. [CrossRef] [PubMed]

24. Makkonen, U.; Hellén, H.; Anttila, P.; Ferm, M. Size distribution and chemical composition of airborne particles in south-eastern Finland during different seasons and wildfire episodes in 2006. *Sci. Total Environ.* **2010**, *408*, 644–651. [CrossRef] [PubMed]

25. Yin, J.; Harrison, R.M. Pragmatic mass closure study for PM$_{1.0}$, PM$_{2.5}$ and PM$_{10}$ at roadside, urban background and rural sites. *Atmos. Environ.* **2008**, *42*, 980–988. [CrossRef]

26. Pérez, N.; Pey, J.; Querol, X.; Alastuey, A.; López, J.M.; Viana, M. Partitioning of major and trace components in PM$_{10}$–PM$_{2.5}$–PM$_1$ at an urban site in Southern Europe. *Atmos. Environ.* **2008**, *42*, 1677–1691. [CrossRef]

27. Chiari, M.; Del Carmine, P.; Lucarelli, F.; Marcazzan, G.; Nava, S.; Paperetti, L.; Prati, P.; Valli, G.; Vecchi, R.; Zucchiatti, A. Atmospheric aerosol characterisation by Ion Beam Analysis techniques: Recent improvements at the Van de Graaff laboratory in Florence. *Nucl. Instrum. Methods Phys. Res. B* **2004**, *219*, 166–170. [CrossRef]

28. Chiari, M.; Lucarelli, F.; Mazzei, F.; Nava, S.; Paperetti, L.; Prati, P.; Valli, G.; Vecchi, R. Characterization of airborne particulate matter in an industrial district near Florence by PIXE and PESA. *X-ray Spectrom.* **2005**, *34*, 323–329. [CrossRef]

29. Lim, S.; Lee, M.; Lee, G.; Kim, S.; Yoon, S.; Kang, K. Ionic and carbonaceous compositions of PM$_{10}$, PM$_{2.5}$ and PM$_{1.0}$ at Gosan ABC Superstation and their ratios as source signature. *Atmos. Chem. Phys.* **2012**, *12*, 2007–2024. [CrossRef]

30. Moreno, T.; Querol, X.; Alastuey, A.; Reche, C.; Cusack, M.; Amato, F.; Pandolfi, M.; Pey, J.; Richards, A.; Prévôt, A.S.H.; et al. Variations in time and space of trace metal aerosol concentrations in urban areas and their surroundings. *Atmos. Chem. Phys.* **2011**, *11*, 9415–9430. [CrossRef]

31. Theodosi, C.; Grivas, G.; Zarmpas, P.; Chaloulakou, A.; Mihalopoulos, N. Mass and chemical composition of size-segregated aerosols (PM_1, $PM_{2.5}$, PM_{10}) over Athens, Greece: Local versus regional sources. *Atmos. Chem. Phys.* **2011**, *11*, 11895–11911. [CrossRef]

32. Speranza, A.; Caggiano, R.; Margiotta, S.; Trippetta, S. A novel approach to comparing simultaneous size-segregated particulate matter (PM) concentration ratios by means of a dedicated triangular diagram using the Agri Valley PM measurements as an example. *Nat. Hazards Earth Syst. Sci.* **2014**, *14*, 2727–2733. [CrossRef]

33. Speranza, A.; Caggiano, R.; Margiotta, S.; Summa, V.; Trippetta, S. A clustering approach based on triangular diagram to study the seasonal variability of simultaneous measurements of PM_{10}, $PM_{2.5}$ and PM_1 mass concentration ratios. *Arab. J. Geosci.* **2016**, *9*, 1–8. [CrossRef]

34. Aitchison, J. The statistical analysis of compositional data (with discussion). *J. R. Stat. Soc. Ser. B Stat. Methodol.* **1982**, *44*, 139–177.

35. Aitchison, J. *The Statistical Analysis of Compositional Data*; Chapman and Hall: London, UK, 1986; 416p, ISBN 0-412-28060-4.

36. Pawlowsky-Glahn, V.; Buccianti, A. Visualization and modeling of sub-populations of compositional data: Statistical methods illustrated by means of geochemical data from fumarolic fluids. *Int. J. Earth Sci.* **2002**, *91*, 357–368. [CrossRef]

37. Martín-Fernández, J.A.; Barceló-Vidal, C.; Pawlowsky-Glahn, V. Dealing with zeros and missing values in compositional datasets using nonparametric imputation. *Math. Geol.* **2003**, *35*, 253–278. [CrossRef]

38. Buccianti, A.; Pawlowsky-Glahn, V. Statistical evaluation of compositional changes in volcanic gas chemistry: A case study. *Stoch. Environ. Res. Risk Assess.* **2006**, *21*, 25–33. [CrossRef]

39. Pawlowsky-Glahn, V.; Buccianti, A. *Compositional Data Analysis: Theory and Applications*; John Wiley & Sons: Chichester, UK, 2011; ISBN 9780470711354.

40. Buccianti, A.; Pawlowsky-Glahn, V. New perspectives on water chemistry and compositional data analysis. *Math. Geol.* **2005**, *37*, 703–727. [CrossRef]

41. Visser, S.; Slowik, J.G.; Furger, M.; Zotter, P.; Bukowiecki, N.; Dressler, R.; Flechsig, U.; Appel, K.; Green, D.C.; Tremper, A.H.; et al. Kerb and urban increment of highly time-resolved trace elements in PM_{10}, $PM_{2.5}$ and $PM_{1.0}$ winter aerosol in London during ClearfLo 2012. *Atmos. Chem. Phys.* **2015**, *15*, 2367–2386. [CrossRef]

42. Putaud, J.P.; Van Dingenen, R.; Raes, F. Submicron aerosol mass balance at urban and semirural sites in the Milan area (Italy). *JGR-Atmospheres* **2002**, *107*, 8198. [CrossRef]

43. Putaud, J.P.; Raes, F.; Van Dingenen, R.; Brüggemann, E.; Facchini, M.C.; Decesari, S.; Fuzzi, S.; Gehrig, R.; Hüglin, C.; Laj, P.; et al. A European aerosol phenomenology—2: Chemical characteristics of particulate matter at kerbside, urban, rural and background sites in Europe. *Atmos. Environ.* **2004**, *38*, 2579–2595. [CrossRef]

44. Colbeck, I.; Nasir, Z.A.; Ahmad, S.; Ali, Z. Exposure to PM_{10}, $PM_{2.5}$, PM_1 and carbon monoxide on roads in Lahore, Pakistan. *Aerosol Air Qual. Res.* **2011**, *11*, 689–695. [CrossRef]

45. Vallius, M.J.; Ruuskanen, J.; Mirme, A.; Pekkanen, J. Concentrations and estimated soot content of PM_1, $PM_{2.5}$, and PM_{10} in a subarctic urban atmosphere. *Environ. Sci. Technol.* **2000**, *34*, 1919–1925. [CrossRef]

46. Lundgren, D.A.; Hlaing, D.N.; Rich, T.A.; Marple, V.A. $PM_{10}/PM_{2.5}/PM_1$ data from a trichotomous sampler. *Aerosol Sci. Technol.* **1996**, *25*, 353–357. [CrossRef]

47. Pawlowsky-Glahn, V.; Egozcue, J.J. Compositional data and their analysis: An introduction. In *Compositional Data Analysis in the Geosciences: From Theory to Practice*; Special Publications; Buccianti, A., Mateu-Figueras, G., Pawlowsky-Glahn, V., Eds.; Geological Society: London, UK, 2006; Volume 264, pp. 1–10.

48. Egozcue, J.J.; Pawlowsky-Glahn, V.; Mateu-Figueras, G.; Barceló-Vidal, C. Isometric logratio transformations for compositional data analysis. *Math. Geol.* **2003**, *35*, 279–300. [CrossRef]

49. Pawlowsky-Glahn, V.; Egozcue, J.J.; Tolosana Delgado, R. Lecture Notes on Compositional Data Analysis. 2007. Available online: http://dugi-doc.udg.edu//handle/10256/297 (accessed on 8 February 2008).

50. Graham, D.J.; Midgley, N.G. TECHNICAL COMMUNICATION-Graphical Representation of Particle Shape using Triangular Diagrams: An Excel Spreadsheet Method. *Earth Surf. Processes Landf.* **2000**, *25*, 1473–1478. [CrossRef]

51. Comas-Cufí, M.; Thió-Henestrosa, S. CoDaPack 2.0: A stand-alone, multi-platform compositional software. In Proceedings of the CoDaWork'11: 4th International Workshop on Compositional Data Analysis, Sant Feliu de Guíxols, Girona, Spain, 10–13 May 2011.

52. R Development Core Team R. *A Language and Environment for Statistical Computing*; R Foundation for Statistical Computing: Vienna, Austria, 2008; ISBN 3-900051-07-0.

53. Buccianti, A.; Pawlowsky-Glahn, V.; Barceló-Vidal, C.; Jarauta-Bragulat, E. Visualization and modeling of natural trends in ternary diagrams: A geochemical case study. In Proceedings of the IAMG, Trondheim, Norway, 6–11 August 1999; Volume 99, pp. 139–144.

54. Martín-Fernández, J.A.; Bren, M.; Barceló-Vidal, C.; Pawlowsky-Glahn, V. A measure of difference for compositional data based on measures of divergence. In Proceedings of the IAMG, Trondheim, Norway, 6–11 August 1999; Volume 99, pp. 211–216.

55. Von Eynatten, H.; Pawlowsky-Glahn, V.; Egozcue, J.J. Understanding perturbation on the simplex: A simple method to better visualize and interpret compositional data in ternary diagrams. *Math. Geol.* **2002**, *34*, 249–257. [CrossRef]

56. Aitchison, J.; Egozcue, J.J. Compositional data analysis: Where are we and where should we be heading? *Math. Geol.* **2005**, *37*, 829–850. [CrossRef]

57. Pawlowsky-Glahn, V.; Egozcue, J.J.; Tolosana-Delgado, R. *Modeling and Analysis of Compositional Data*; John Wiley & Sons: Chichester, UK, 2015; ISBN 9781118443064.

58. Stephens, M.A. EDF statistics for goodness of fit and some comparisons. *J. Am. Stat. Assoc.* **1974**, *69*, 730–737. [CrossRef]

59. Aitchison, J. A Concise Guide to Compositional Data Analysis. In *2nd Compositional Data Analysis Workshop—CoDaWork'05*; Universitat de Girona: Girona, Spain, 2005; Available online: http://ima.udg.edu/Activitats/CoDaWork05/A_concise_guide_to_compositional_data_analysis.pdf (accessed on 19 October 2015).

60. Harrison, R.M.; Yin, J.; Mark, D.; Stedman, J.; Appleby, R.S.; Booker, J.; Moorcroft, S. Studies of the coarse particle (2.5–10 μm) component in UK urban atmospheres. *Atmos. Environ.* **2001**, *35*, 3667–3679. [CrossRef]

61. Jones, A.M.; Harrison, R.M. Assessment of natural components of PM_{10} at UK urban and rural sites. *Atmos. Environ.* **2006**, *40*, 7733–7741. [CrossRef]

Article

The New Beamline LISA at ESRF: Performances and Perspectives for Earth and Environmental Sciences

Alessandro Puri *, Giovanni Orazio Lepore and Francesco d'Acapito

CNR-IOM-OGG c/o ESRF—The European Synchrotron, LISA CRG, 71 Avenue des Martyrs, CS 40220, Grenoble F-38043, France; lepore@esrf.fr (G.O.L.); dacapito@esrf.fr (F.d.)
* Correspondence: puri@esrf.fr

Received: 29 November 2018; Accepted: 12 January 2019; Published: 15 January 2019

Abstract: LISA (Linea Italiana per la Spettroscopia di Assorbimento di raggi X) is the new Italian Collaborating Research Group (CRG) beamline at the European Synchrotron Radiation Facility (ESRF) dedicated to X-ray absorption spectroscopy (XAS). The beamline covers a wide energy range, $4 < E < 90$ keV, which offers the possibility for probe the K and L edges of elements that are heavier than Ca. A liquid He/N_2 cryostat and a compact furnace are available for measurements in a wide temperature range (10–1000 K), allowing for in situ chemical treatments and measurements under a controlled atmosphere. The sub-millimetric beam size, the high photon flux provided, and the X-ray fluorescence detectors available (HP-Ge, SDD) allow for the study of liquid and highly diluted samples. Trace elements in geological or environmental samples can be analyzed, even for very small sample areas, gaining information on oxidation states and host phases.

Keywords: XAS spectroscopy; synchrotron radiation; environment

1. Introduction

LISA (Linea Italiana per la Spettroscopia di Assorbimento di raggi X) [1,2], the new Italian beamline at the European Synchrotron Radiation Facility (ESRF), is the result of the refurbishment of the former GILDA (General purpose Italian beam Line for Diffraction and Absorption) [3], and it has been open to users since April 2018. The beamline has been designed and optimized for X-ray absorption spectroscopy (XAS).

All the optics were completely renewed, and are fully compatible with the single bending magnet source of the future EBS ring. The X-ray optics consist of two mirrors and one monochromator. The first cylindrical mirror (positioned before the monochromator) collimates the beam in the vertical plane. The double crystal monochromator (DCM) allows the access of a wide energy range (4–90 keV), which offers the possibility to probe the K- and L- edges of most of the elements heavier than calcium (Ca). The DCM contains two crystal pairs: Si(311) for experiments needing high energy resolution, and Si(111) for experiments needing high photon flux. The second mirror (toroidal, placed after the DCM) focalizes the beam about 17 m downstream on a roughly circular spot of diameter $\leq$200 μm. Both mirrors are made of a single crystal Si substrate with two optical regions: one exposing the silicon, and the other coated with Pt. The mirrors operate at a working angle of 2 mrads, with resulting energy cut-off of about 15 and 40 keV for the Si and Pt stripes, respectively. In order to reach energies above 40 keV the mirrors are moved out of the beam. In addition, a pair of Pt coated flat mirrors working at 8 mrads (cutoff $\approx$ 12 keV), located after the second mirror, can be used for harmonics rejection at low energies (4–6 keV). The small spot size and the beam stability during energy scans [2] (beam movement <10% of the beam size over an angular range of 30° for both crystals pairs) allow for the study of samples of reduced dimensions, like small crystals and fibers, or to obtain elemental mappings of the sample with a spatial resolution of down to ~100×65 μm^2.

The new optics design provides a photon flux of 10^{11} ph/s on the sample with the Si(111) crystal pair; this, together with the fluorescence detectors available, a 13-elements ORTEC IGLET-ARRAY-UPGRADE-S and a 12-elements ORTEC C-TRAIN-1200-S High Purity-Ge (HP-Ge), plus a 4-channel Silicon Drift Detector (SDD) ARDESIA [4], allows for the analysis of trace elements in geological or environmental samples, and the study of highly diluted and liquid samples. The detection limit depends on the type of sample (powder, thin film, liquid...) and on the matrix absorption. The estimated value is 10 ppm in atoms, and 1/10 of a monolayer for films.

After the optimization of the beamline optical configuration, the resulting energy resolution, evaluated from the rocking curves Full Width at Half Maximum (FWHM) of the two crystal sets measured in a wide energy range (5–40 keV), is $\Delta E/E = 10^{-4}$–10^{-5} [2], depending on the crystal pair, which is in excellent agreement with theoretical calculations [5].

These features open the possibility of addressing many environmental issues, such as investigating trace elements cycles in the environment, the chemical activities of environmental pollutants, elements speciation, and the characterization of natural/anthropogenic complex matrixes that are not suitable for standard chemical and structural analyses, despite being of crucial importance in environmental sciences.

In this paper, we report on some recent research performed at LISA, pointing out the new beamline possibilities and the work done at the beamline over the last few years in the environmental sciences.

2. LISA Experimental Hutches

LISA has two experimental hutches, the principal one (EH2), approximately centered on the focal position of the second mirror, while the other (EH1) is a few meters upstream.

The principal experimental hutch EH2 (Figure 1) is dedicated to experiments needing a focused beam, mainly Extended X-ray Absorption Fine Structure (EXAFS) in fluorescence mode, Grazing Incidence X-ray Absorption Spectroscopy (GIXAS), reflectivity EXAFS (REFLEXAFS), or experiments with diluted or small samples. The experimental apparatuses (vacuum chambers, ion chambers, detectors) are installed on the new EH2 bench, and can be shifted along the beam path, via sliding plates, in order to accommodate additional instrumentation, and to adapt to the user's needs.

Figure 1. The LISA (Linea Italiana per la Spettroscopia di Assorbimento di raggi X) new experimental hutch, EH2.

Figure 1 shows a typical setup: the main vacuum chamber is equipped with a liquid He/N$_2$ cold finger cryostat (10–300 K) and it is preceded and followed by the ion chambers I0 and I1, respectively. Samples are aligned with motorized vertical and horizontal translations, plus motorized rotations along the vertical axis. Downstream of I1, a small vacuum chamber and a further ion chamber I2 are used to collect reference compound spectra. The fluorescence detectors are mounted on the machine wall side of the bench, and can be approached or moved away from the sample with a motorized translation. EH1 is equipped with a vacuum chamber and two ionization chambers. This hutch is dedicated to measurements in transmission mode with an mm-sized beam (typically 2×2 mm^2), as it is far from the focal point of the second mirror. Furthermore, the beamline is equipped with a compact furnace for measurements at high temperature (up to 1000 K), under a controlled atmosphere, and is thus suitable for in-situ chemical treatments.

Energy scans are performed in step mode. Users can adjust the energy mesh for each region of the XAS spectrum. The dead time is 1.5 s per point in transmission mode, and slightly higher in fluorescence mode, due to the detector readout. We would like to stress the fact that the main goal of the beamline is to obtain high-quality low-noise measurements with very good energy stability.

3. XAS Spectra of a Single Riebeckite Fiber

As an example of the performances of the new LISA beamline, in Figure 2, we report the Fe K-edge XAS spectra of a single 500 µm sized fiber of synthetic riebeckite ($Na_2Fe^{2+}_3Fe^{3+}_2(Si_8O_{22})(OH)_2$). The fibrous forms of riebeckite are known as crocidolite, and are one of the six types of asbestos. They represent a serious issue for health, due to their carcinogen effects, which are thought to be involved in malignant mesothelioma and lung cancer. However, the mechanism by which asbestos and other mineral fibers cause cancer is still poorly understood. The iron contained in the fiber is suspected to play a prominent role in the process.

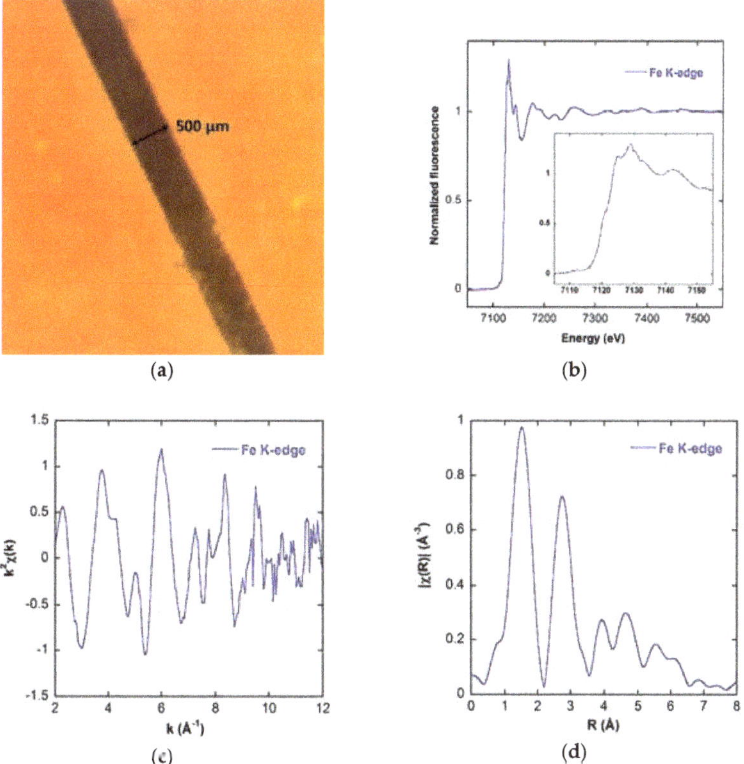

Figure 2. (a) Photograph of the riebeckite fiber embedded in kapton; (b) Fe K-edge Extended X-ray Absorption Fine Structure (EXAFS) of the fiber, the inset shows the X-ray Absorption Near-Edge Structure (XANES) region; (c) k^2-weighted EXAFS spectrum; (d) k^2-weighted EXAFS Fourier Transform magnitude.

The spectra have been collected in fluorescence mode using the ORTEC 12-elements HP-Ge detector mounted behind the measurement chamber (Figure 1), and the Si(311) crystal pair. The beam was focalized in a 200 µm spot and the fiber, rotated 45° with respect to the incoming X-ray beam, was oriented with the long axis, i.e., the *c*-axis, parallel to the radiation polarization plane, allowing a single fiber embedded in kapton to be probed. A time period of 90 min were necessary to acquire a complete spectrum, with 10 s counting time per point.

A good signal-to-noise ratio was obtained up to 10 Å^{-1} in k, allowing for the resolution of several coordination shells around the Fe atom. The EXAFS data analysis is in progress to better clarify the structural environment of iron.

In Figure 3, we compare the pre-edge of the single fiber with that of a powdered sample of riebeckite. In the fit, we used Gaussian curves for each component, left free to vary the energy position in a range of ±0.75 eV from the initial value. Their FWHM was constrained to be constant at ~1.5 eV (the core-hole of Fe). The incoming energy resolution was ~0.3 eV [6,7].

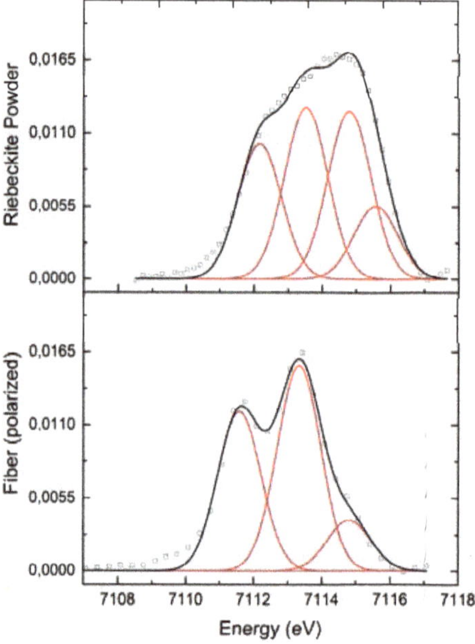

Figure 3. Comparison between the pre-edge of a powdered sample of riebeckite (**top**) and that of a 500 µm thick fiber of riebeckite (**bottom**).

The spectra show clear differences: for the single fiber oriented parallel to the beam polarization the pre-edge is sharper, and two features are easily visible, while it is broader for the powder sample, where four contributions are needed in the fit, probably due to the multi-orientation of the grains. This evidences the importance of having a small polarized beam in these kinds of experiments.

4. Chemical Activities of Environmental Pollutants in Aerosols Stored in Snow and Ice-Cores from Western China and the Arctic Atmosphere

A long-term project on the investigation of snow and ice samples retrieved from glaciers located around the Tibetan plateau and Antarctica is currently being carried out at LISA, to shed light on the role played by iron in climatic and environmental systems, in particular regarding their relationship with atmospheric dust and the aerosol cycle.

Snow and ice deposited on the glaciers represent an invaluable archive of climatic and environmental information, in particular, with respect to the atmosphere. The Tibetan region and its glaciers are surrounded by the most populated and industrialized areas. On the other hand, Antarctica, due to its remoteness with respect to human activities, can be considered as a global background, which is extremely useful for assessing the trends and mechanisms that are modifying the global biogeochemical cycles of many elements.

A selection of Fe K-edge XANES and EXAFS spectra of soil samples is presented in Figure 4; the spectra have been collected in fluorescence mode by using the same geometrical configuration described in Section 3. A SDD single channel detector Vortex EM from the ESRF instrument pool and the Si(111) crystal pair were used. Each spectra was acquired in 50 min (5 s counting time per point), and in some cases for the most diluted samples several spectra were averaged for the final spectrum. In this way, although the statistic was not exactly the same, the noise resulted to be sufficiently low to allow a reliable data analysis. The origin, type, and average Fe concentration of the samples are indicated in Table 1.

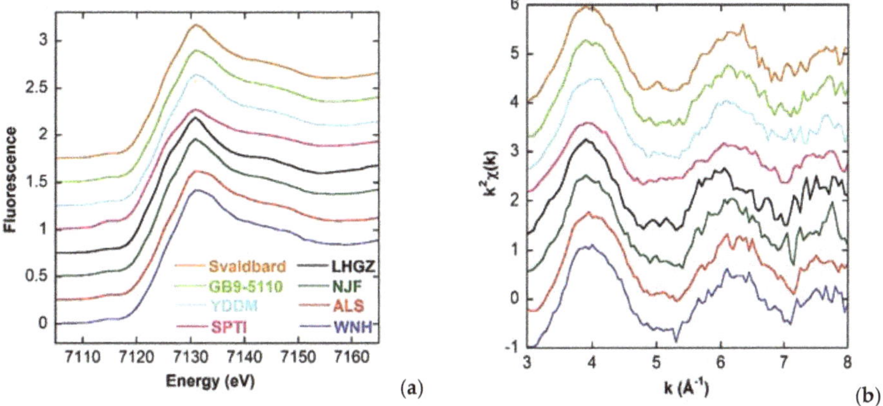

Figure 4. Comparison of several spectra at the Fe K-edge from different samples: (a) XANES region; (b) k^2-weighted EXAFS signal. The curves are shifted vertically for clarity.

Table 1. Origins of samples presented in Figure 3.

Name	Type	Average Fe Concentration (mg/kg)	Location
Svalbard	sand	1.99×10^4 [8]	Svalbard
GB9-5110	sand	2.50×10^3	Inner Mongolia
YDDM	sand	2.50×10^3	Near Dunhuang City
SPTI	sand	2.50×10^3	Near Tengger desert
LHGZ	sand	2.50×10^3	Laohugou glacier No. 12 station (Western Qilian Mountain)
NJF	sand	2.50×10^3	Zhongshan Station (Antarctica)
ALS	loess	1.03×10^1	Wulatehouqi (inner Mongolia)
WNH	loess	1.03×10^1	Chinese loess plateau (near Xian city)

The spectra show marked differences: sample SPTI shows a more intense pre-edge peak at 7115 eV, a broad hump at 7126 eV, and a less pronounced white line at 7131 eV; sample LHGZ shows two more pronounced features at 7122 eV and 7126 eV, and a sharper white line at 7131 eV, which on the contrary, is broader for the ALS and WNH samples; these last two samples show a broad hump at 7125 eV. These differences point out the variability in Fe oxidation, coordination, and mineralogy in samples of different types and provenances.

5. Recent Research in Environmental Science

LISA and the former GILDA have allowed for a broad range of Earth and environmental issues to be addressed recently. Besides the long-term project on the study of snow and ice samples from the Tibetan plateau and Antarctica regions discussed above, a similar study of iron speciation in Arctic snow and soil samples from the Spitzbergen region (Svalbard) is under investigation. A great deal of effort is being devoted to the study of contaminants in the environment, especially regarding the particulate air pollution, along with the distribution of heavy metals or toxic elements in water, soil, plants, and to the speciation and reaction pathways of metals and other environmentally relevant elements in minerals and rocks.

Among the published works based on measurements performed at the beamline in the last few years, we point out the investigation of Tl contamination of the drinkable water from the public distribution system of Valdicastello Carducci-Pietrasanta (northern Tuscany, Italy) [9]; the study of As contamination of floodplain soils and groundwater from shallow aquifers [10,11]; the study of Fe chemical environment in asbestos and erionite mineral fibers [12]; the investigation of selenium (Se) and lead (Pb) accumulation, distribution, and speciation in edible plants [13,14]; the Fe speciation in urban particulate matter and industrial quartz dusts [15,16]; the investigation of amorphous zinc (Zn) biomineralization ("white mud"), occurring at Naracauli stream, Sardinia [17]; the Sb speciation in brake linings, brake pad wear residues, road dust, and atmospheric particulate matter [18]; the interaction of As-rich natural pyrite with aqueous selenite [Se (IV)] [19], and As speciation in travertines [20].

6. Conclusions

The new Italian CRG beamline LISA has completed its construction, and it has been open to users since April 2018. The high photon flux (10^{11} ph/s on the sample with Si(111) crystal pair), the reduced beam size (less than 200 μm beam spot), and the possibility to work in fluorescence mode, provide a valuable tool for earth and environmental scientists. Highly diluted samples and trace elements in host matrices can now be analyzed with unprecedented accuracy and precision compared to the previous GILDA beamline.

For more information and the latest news about the beamline, please visit the LISA webpage: http://www.esrf.eu/UsersAndScience/Experiments/CRG/BM08.

Author Contributions: Conceptualization, G.O.L.; Data curation, A.P.; Supervision, F.d.; Writing—original draft, A.P.; Writing—review & editing, G.O.L. and F.d.

Funding: This research received no external funding.

Acknowledgments: We acknowledge Giancarlo Della Ventura (Roma Tre University) for the fiber sample presented in Section 2, Federico Galdenzi (Roma Tre University) for Figure 3 and the fit presented in Section 2. We acknowledge Augusto Marcelli (INFN-Laboratori Nazionali di Frascati), Giannantonio Cibin (Diamond Light Source), Giovanni Baccolo (Università degli Studi di Milano-Bicocca), Valter Maggi (Università degli Studi di Milano-Bicocca), and Du Zhiheng (Chinese Academy of Sciences), for the data presented in Section 3.

Conflicts of Interest: The authors declare no conflict of interest.

References

1. d'Acapito, F.; Trapananti, A.; Puri, A. LISA: The Italian CRG beamline for X-ray Absorption Spectroscopy at ESRF. *J. Phys. Conf. Ser.* **2016**, *712*, 012021. [CrossRef]

2. d'Acapito, F.; Lepore, G.O.; Puri, A.; Laloni, A.; La Mannna, F.; Dettona, E.; De Luisa, A.; Martin, A. The LISA beamline at ESRF. *J. Synchrotron Radiat.* **2019**, in press.

3. d'Acapito, F.; Trapananti, A.; Torrengo, S.; Mobilio, S. X-ray Absorption Spectroscopy: The Italian beamline GILDA of the ESRF. *Notiziario Neutroni e Luce di Sincrotrone* **2014**, *19*, 14–23.

4. Bellotti, G.; Butt, A.D.; Carminati, M.; Fiorini, C.; Bombelli, L.; Borghi, G.; Piemonte, C.; Zorzi, N.; Balerna, A. ARDESIA Detection Module: A Four-Channel Array of SDDs for Mcps X-Ray Spectroscopy in Synchrotron Radiation Applications. *IEEE Trans. Nucl. Sci.* **2018**, *65*, 1355–1364. [CrossRef]

5. Ishikawa, T.; Tamasaku, K.; Yabashi, M. High-resolution X-ray monochromators. *Nucl. Instrum. Methods Phys. Res. A* **2005**, *547*, 42–49. [CrossRef]

6. Galdenzi, F.; Della Ventura, G.; Cibin, G.; Macis, S.; Marcelli, A. Accurate Fe3+/Fetot ratio from XAS spectra at the Fe K-edge. *Radiat. Phys. Chem.* **2018**, in press. [CrossRef]

7. Della Ventura, G.; Galdenzi, F.; Cibin, G.; Oberti, R.; Xu, W.; Macis, S.; Marcelli, A. Iron oxidation dynamics *vs.* temperature of synthetic potassic-ferro-richterite: A XANES investigation. *Phys. Chem. Chem. Phys.* **2018**, *20*, 21764–21771. [CrossRef] [PubMed]

8. Zhan, J.; Gao, Y.; Li, W.; Chen, L.; Lin, H.; Lin, Q. Effects of ship emissions on summertime aerosols at Ny–Alesund in the Arctic. *Atmos. Pollut. Res.* **2014**, *5*, 500–510. [CrossRef]

9. Biagioni, C.; D'Orazio, M.; Lepore, G.O.; d'Acapito, F.; Vezzoni, S. Thallium-rich rust scales in drinkable water distribution systems: A case study from northern Tuscany, Italy. *Sci. Total Environ.* **2017**, *587–588*, 491–501. [CrossRef] [PubMed]

10. Parsons, C.T.; Couture, R.-M.; Omoregie, E.O.; Bardelli, F.; Greneche, J.-M.; Roman-Ross, G.; Charlet, L. The impact of oscillating redox conditions: Arsenic immobilisation in contaminated calcareous floodplain soils. *Environ. Pollut.* **2013**, *178*, 254–263. [CrossRef] [PubMed]

11. Phan, V.T.H.; Bonnet, T.; Garambois, S.; Tisserand, D.; Bardelli, F.; Bernier-Latmani, R.; Charlet, L. Arsenic in Shallow Aquifers Linked to the Electrical Ground Conductivity: The Mekong Delta Source Example. *Geosci. Res.* **2017**, *2*, 180–195.

12. Pollastri, S.; D'Acapito, F.; Trapananti, A.; Colantoni, I.; Andreozzi, G.B.; Gualtieri, A.F. The chemical environment of iron in mineral fibres. A combined X-ray absorption and Mössbauer spectroscopic study. *J. Hazard. Mater.* **2015**, *298*, 282–293. [CrossRef] [PubMed]

13. Eiche, E.; Bardelli, F.; Nothstein, A.K.; Charlet, L.; Göttlicher, J.; Steininger, R.; Dhillon, K.S.; Sadana, U.S. Selenium distribution and speciation in plant parts of wheat (Triticum aestivum) and Indian mustard (Brassica juncea) from a seleniferous area of Punjab, India. *Sci. Total. Environ.* **2015**, *505*, 952–961. [CrossRef] [PubMed]

14. Massaccesi, L.; Meneghini, C.; Comaschi, T.; D'Amato, R.; Onofri, A.; Businelli, D. Ligands involved in Pb immobilization and transport in lettuce, radish, tomato and Italian ryegrass. *J. Plant Nutr. Soil Sci.* **2014**, *177*, 766–774. [CrossRef]

15. d'Acapito, F.; Mazziotti Tagliani, S.; Di Benedetto, F.; Gianfagna, A. Local order and valence state of Fe in urban suspended particulate matter. *Atmos. Environ.* **2014**, *99*, 582–586. [CrossRef]

16. Di Benedetto, F.; D'Acapito, F.; Capacci, F.; Fornaciai, G.; Innocenti, M.; Montegrossi, G.; Oberhauser, W.; Pardi, L.A.; Romanelli, M. Variability of the health effects of crystalline silica: Fe speciation in industrial quartz reagents and suspended dusts—Insights from XAS spectroscopy. *Phys. Chem. Miner.* **2014**, *41*, 215–225. [CrossRef]

17. Medas, D.; Lattanzi, P.; Podda, F.; Meneghini, C.; Trapananti, A.; Sprocati, A.; Casu, M.A.; Musu, E.; De Giudici, G. The amorphous Zn biomineralization at Naracauli stream, Sardinia: Electron microscopy and X-ray absorption spectroscopy. *Environ. Sci. Pollut. Res. Int.* **2014**, *21*, 6775–6782. [CrossRef] [PubMed]

18. Varrica, D.; Bardelli, F.; Dongarrà, G.; Tamburo, E. Speciation of Sb in airborne particulate matter, vehicle brake linings, and brake pad wear residues. *Atmos. Environ.* **2013**, *64*, 18–24. [CrossRef]

19. Kang, M.; Bardelli, F.; Charlet, L.; Géhin, A.; Shchukarev, A.; Chen, F.; Morel, M.-C.; Ma, B.; Liu, C. Redox reaction of aqueous selenite with As-rich pyrite from Jiguanshan ore mine (China): Reaction products and pathway. *Appl. Geochem.* **2014**, *47*, 130–140. [CrossRef]

20. Winkel, L.H.E.; Casentini, B.; Bardelli, F.; Voegelin, A.; Nikolaidis, N.P.; Charlet, L. Speciation of arsenic in Greek travertines: Co-precipitation of arsenate with calcite. *Geochim. Cosmochim. Acta* **2013**, *106*, 99–110. [CrossRef]

Review

Post-Depositional Biodegradation Processes of Pollutants on Glacier Surfaces

Francesca Pittino [1] , **Roberto Ambrosini** [2], **Roberto S. Azzoni** [2] , **Guglielmina A. Diolaiuti** [2], **Sara Villa** [1], **Isabella Gandolfi** [1] **and Andrea Franzetti** [1,*]

[1]   Department of Earth and Environmental Sciences, University of Milano Bicocca, Piazza della Scienza, 1 20126 Milano, Italy; f.pittino@campus.unimib.it (F.P.); sara.villa@unimib.it (S.V.); isabella.gandolfi@unimib.it (I.G.)

[2]   Department of Environmental Science and Policy, University of Milan, via Celoria, 2 20133 Milano, Italy; roberto.ambrosini@unimi.it (R.A.); robertosergio.azzoni@unimi.it (R.S.A.); guglielmina.diolaiuti@unimi.it (G.A.D.)

*   Correspondence: andrea.franzetti@unimib.it; Tel.: +39-02-64482927

Received: 14 July 2018; Accepted: 9 August 2018; Published: 11 August 2018

Abstract: Glaciers are important fresh-water reservoirs for our planet. Although they are often located at high elevations or in remote areas, glacial ecosystems are not pristine, as many pollutants can undergo long-range atmospheric transport and be deposited on glacier surface, where they can be stored for long periods of time, and then be released into the down-valley ecosystems. Understanding the dynamics of these pollutants in glaciers is therefore important for assessing their environmental fate. To this aim, it is important to study cryoconite holes, small ponds filled with water and with a layer of sediment, the cryoconite, at the bottom, which occur on the surface of most glaciers. Indeed, these environments are hotspots of biodiversity on glacier surface as they host metabolically active bacterial communities that include generalist taxa able to degrade pollutants. In this work, we aim to review the studies that have already investigated pollutant (e.g., chlorpyrifos and polychlorinated-biphenyls (PCBs)) degradation in cryoconite holes and other supraglacial environmental matrices. These studies have revealed that bacteria play a significant role in pollutant degradation in these habitats and can be positively selected in contaminated environments. We will also provide indication for future research in this field.

Keywords: cryoconite; POPs; microbiology; long-range transport; cryosphere; contaminants; bacteria

---

1. Introduction

For more than 40 years, we have known that inorganic and organic pollutants are present in cold remote areas, such as polar and mountain regions, far from their emission sources. Early studies were conducted in Arctic regions and reported high concentrations of radionuclides in layers corresponding to nuclear test periods [1] and the presence of chlorinated compounds in Arctic and Antarctic regions [2,3]. The organic contaminants in these remote areas are referred to as persistent organic pollutants (POPs), which consist of several groups of chemicals with similar structures and physical-chemical properties that are extensively used worldwide in agriculture (pesticides), industrial and health applications. Recent studies have revealed that alpine environments are affected by the presence of POPs [4–6]. These chemicals are ubiquitous, show long-range transport potential, and many of them are hydrophobic. In addition, they bio-accumulate in organisms via respiration, dermal contact, and through diet [7]. Some classes of contaminants comprised in the POPs are polychlorinated-biphenyls (PCBs), -dioxins (PCDDs), -furans (PCDFs), polybrominated-diphenyl ethers (PBDEs), -biphenyls (PBBs), perfluorinated compounds (PFCs) and other halogenated hydrocarbons, used as pesticides [8].

PCBs are one of the classes of compounds that have been studied in the context of pollutant biodegradation on glaciers [9]. They are toxic chemicals that the chemical treaty is trying to abolish by 2025, because of their persistence and difficult degradation. Their use is very heterogeneous (from the industry to agriculture) [10]. PCB biodegradation can occur either through anaerobic reductive dechlorination (mostly in soil and sediment) or through aerobic oxidative degradation (preferably in water) and a wide range of different bacterial genera are able to perform these degradative pathways [10,11]. PCB degradation rate, measured in lab experiments, varies according to the number of the chlorine atoms, slowing down as the number of Cl^- atoms increases. This is the reason why the range of the degradation rate varied from approximately 90% to approximately 20% [12].

Atmospheric medium- and long-range transport has been identified as a major source of semivolatile and persistent organic contaminants in polar and alpine ecosystems. Cold areas act as condensers [13], interfering with the atmospheric transport and global cycling of semi volatile organic compounds (SVOCs) [14–18], thus promoting the scavenging of these molecules from the atmosphere [19]. Indeed, PCB concentrations in snow increase with altitude, especially for more-volatile, less-chlorinated di- and tri-chlorobiphenyl congeners [4]. High altitude environments may also be more prone than polar ones to contamination by those compounds that are not suitable for long latitudinal transport because mountain ranges can be relatively close to source sites of contaminants [20].

A relevant fraction of the contaminants reaching cold areas is deposited on glaciers, where pollutants undergo post-depositional processes of partitioning among different environmental matrices (e.g., snow, ice, water, interstitial atmospheric gases and supraglacial sediments) and alteration processes. Post-depositional alteration consists of both physical-chemical processes, such as photodegradation, hydrolysis, revolatilization [21,22], and biological ones, particularly biodegradation [6]. The balance of these partition and alteration processes defines the amount of pollutants that can either enter the food-chain or be released into the environment by meltwater, which, in turn, can impact the downstream environments [23,24]. The current and foreseen increase of glacier retreat and melting will therefore likely lead to an increase of the release of pollutants from glaciers. Indeed, it has been already reported that high retreat episodes of glaciers correspond to high fluxes of pollutants in the melting water [25]. A recent model showed that the glacier melting due to global warming will lead to an earlier and more concentrated release of pollutants (e.g., PCBs, DDT, PCDD/Fs) stored in glacier bodies, than if the climate were stable [26].

Among post-depositional alteration processes that lead to the net reduction of the contaminant mass on the glacier, the biodegradation of organic molecules by microorganisms has been rather overlooked so far [6]. However, it is well known that microorganisms inhabit glaciers and have the metabolic abilities to degrade complex organic compounds even at low temperatures [27]. Among the glacial environments, supraglacial ones are the most biodiverse, and host rich bacterial communities [28]. In these environments, the cryoconite, a wind-borne fine debris deposited on glacier surfaces, represents a potential sink for organic and inorganic pollutants because of its high content of organic matter [29]. When heated by solar radiation, the cryoconite can promote ice melting and form small ponds filled by meltwater, called cryoconite holes, which are considered the most microbiologically active supraglacial habitats. Indeed, different studies report bacterial abundance in cryoconite in the order of 10^8–10^9 cells·g^{-1} [30,31], similar to these in other type of soils (e.g., agricultural soils, marsh or mountain soils) [32]. Cryoconite holes therefore represent ideal environments to investigate the processes that determine pollutant accumulation and degradation on glaciers [29].

In this work, we review the current literature about the accumulation of contaminants in the cryoconite and other supraglacial matrices and the microbiological processes affecting their fate, and we provide suggestions and direction for future research.

2. Accumulation of Pollutants and Microbiological Response in Cryoconite

Cryoconite is a granular sediment found on glacier surfaces. Owing to its dark color, cryoconite absorbs solar radiation and promotes the formation of quasi-cylindrical holes, called cryoconite holes, through differential ablation rates compared to the surrounding ice [33]. Cryoconite is composed of both inorganic and organic material. The former is mainly composed by mineral fragments, often dominated by phyllosilicate, tectosilicate and quartz with differences due to the geological source of the debris [33]; the latter is composed of a wide variety of living and dead microscopic organisms, the products of their autotrophic activity, and allochthonous organic material of wind-borne particles [34]. Cryoconite in holes has higher concentration of organic matter and inorganic nutrients (nitrogen, phosphorus...) than the surrounding ice and supraglacial sediments [35,36]. This characteristic leads, on the one hand, to the incorporation of hydrophobic compounds such as POPs into the cryoconite organic matter [6] and, on the other hand, it promotes the presence of biodiverse and active microbial communities [37], which can biodegrade the contaminants.

In the last five years, some studies investigated the origin and the accumulation of pollutants in cryoconite and their post-depositional fate, including the biodegradation processes. Polycyclic Aromatic Hydrocarbons (PAHs) are among these pollutants. They are found in the environment as consequence of combustion or processing of hydrocarbon fuels. Their biodegradation can occur in different environments thanks to bacteria, fungi or algae and the efficiency depends on the number of benzene rings [38]. PAHs are known to be better biodegraded in water, but Kuppusamy and colleagues showed new consortia of bacteria able to perform PAH degradation in soil with good performances [39].

Li and colleagues identified 15 PAHs containing 3–7 rings in 61 cryoconite samples collected from seven glaciers on the Tibetan Plateau (TP) [40]. The average concentration of total PAHs in cryoconite samples was in the range of 6.67–3906.66 ng·g^{-1} dry weight. The highest average total PAH concentration was found in the southeastern TP, followed by the northern TP. The central TP contained the lowest number of PAHs. Moreover, correlation analysis showed that total organic carbon (TOC) and grain size were only minor determinants of the accumulation of PAHs in cryoconite of the TP. Factor analysis and diagnostic ratios indicated that PAHs were produced mainly from the incomplete combustion of coal, fossil fuels and biomasses. The exhaust gas of locomotives also contributed to the accumulation of PAHs on glaciers. Toxicity Equivalent Quantity (TEQ) of cryoconite was calculated for all the glaciers and results showed that cryoconite had a low biological risk regarding PAHs in all the investigated glaciers, except for YL Snow Mountain, which is a touristic area where shuttle vehicles are widely used and five-seven ring PAHs accounted for more than half of total PAHs. Overall results showed that long-range atmospheric transport was the main source of PAHs deposited on glaciers.

Similarly, Dong and colleagues reported that the higher ratio of anthropogenic particles in the southern TP is likely caused by atmospheric pollutant transport from southern Asia, whereas cryoconite in the northern locations of the TP, containing higher dust and salt particle ratio, is influenced by the large deserts in central Asia [41]. Therefore, the transport and deposition of cryoconite is significant for understanding the regional atmospheric environment and circulation. A large amount of material such as biological particles, NaCl, and mixed cation sulfate particles, was also found in the cryoconite, implying that, in addition to dust and black carbon, many types of light absorbing impurities combined could influence the glacier albedo change and enhance ice melting in the mountain glaciers of the TP.

The anthropogenic impact on cryoconite was also studied in the Alps and Arctic where high concentrations of radionuclides were found in two independent studies [42,43]. Both studies attributed the origin of these contaminants to long-range transport and anthropogenic activities. The extremely high concentration of metal and radioactive compounds was explained by the capability of extracellular substances excreted by microorganisms to bind these compounds and remove them from the meltwater. A recent study [44] investigated anthropogenic radionuclides in cryoconite from the Adishi glacier (Georgia) and found activity concentrations varying from 0.37 ± 0.04 Bq·Kg^{-1} for ^{238}Pu (^{238}Pu activity concentrations in the first centimeter of soil in an undisturbed area in Korea was in the range of

0.006–0.062 Bq·Kg^{-1} [45]) to a maximum of 4940 $\pm$ 610 for ^{137}Cs (^{137}Cs activity concentration detected after Fukushima nuclear accident at soil depth of 0.5–1 cm was 5610 $\pm$ 40.8 Bq·Kg^{-1} [46]).

Interactions between radionuclides and bacteria can be very different: among them there are, noteworthy, biotransformation, bioprecipitation and also biosorption, depending on the characteristics of the radionuclides [47]. These compounds can be naturally present in the environment or due to human activity (e.g., nuclear weapons, uranium mining and milling, commercial fuel reprocessing) [48].

Due to the limited knowledge about glacial biology, how these high amounts of radionuclides and heavy metals affect biota and how these contaminants are transferred along food web remain open questions to be addressed in the future investigations.

To the best of our knowledge, only six studies characterized the microbial communities and their metabolisms associated to the contamination, besides addressing the presence of pollutants in cryoconite and other supraglacial environments

The metabolic potential of the culturable fraction of cryoconite bacteria toward pollutants was pioneeringly investigated by Margesin in 2002. In this study they estimated the abundance of bacteria able to grow on natural and anthropogenic recalcitrant substrates [49]. Results showed that cold-adapted bacteria were able to degrade carbohydrates (starch), fats (tributyrin), mineral oil hydrocarbons (diesel oil) and PAHs as sole carbon sources. (Table 1, Figure 1).

Table 1. Summary of the published studies addressing both the presence of pollutants and the microbial processes related to them in cryoconite and other supraglacial environmental matrices.

Geographic Area	Contaminants	Matrices	Reference
Italian Alps	Chlorpyrifos (CPF)	cryoconite	Ferrario et al., 2017 [41]
Antarctica	Synthetic oil	surface ice and sediments	Jarula et al., 2009 [38]
Greenland	2,4-Dichlorophenoxyacetic Acid	surface ice	Stibal et al., 2012 [40]
Svalbard	Mercury	snow (above soil)	Larose et al., 2013 [35]
Austrian Alps	PCBs	cryoconite	Weiland-Bräuer et al., 2017 [39]
Greenland	PAHs, PCBs, mercury, lead	cryoconite	Hauptmann et al., 2017 [44]

Larose and colleagues [50] studied the impact of mercury, whose concentration is increasing in the Arctic food web, on snow microbial communities. Despite the investigated snowpack was not supraglacial, but consisted of seasonal snow accumulated above the soil, the study provides important insights into the response of cold adapted communities to inorganic contaminants.

Mercury can be subjected to long-range atmospheric transport and can persist in the air for at maximum one year: enough to reach also the most remote area of the planet [51]. Hg can be found in different forms: elemental Hg (Hg$^0_{(aq)}$) that is volatile but not reactive, mercuric species (Hg(II)) and organic mercury (e.g., methyl-mercury, that is bioaccumulative) [51]. Bacteria can play an important role reducing the soluble form Hg(II) to the precipitating Hg(0) and it has been demonstrated they can perform this reaction with better results at Hg(II) concentrations lower than 6 mg/L in water [52].

In the study by Larose et al. [50] both forms of mercury resulted to influence the structure of the snow microbial communities. Indeed, the microbiological analyses revealed the presence of mercury-resistant taxa in Hg contaminated snow and, at molecular level, the abundance of the *mer*A gene, which confers resistance to the inorganic form of Hg, was positively correlated with Hg concentration. *mer*A codes for the mercuric reductase enzyme which catalyzes the reduction from the water soluble Hg(II) to the volatile Hg(0) form. The presence of this metabolic activity might also explain the observed reemission in the atmosphere of the Hg entrapped in the snow [53]. Further studies are needed regarding mercury in glacial ecosystems, especially regarding Me–Hg that can be bioaccumulated along the trophic chain [51]. Indeed Larose et al., found a negative correlation between bioavailable Hg and Me–Hg indicating a probable Hg biotic methylation [50]; in fact, bacteria can both cleave the bond or be responsible of methylation especially in anaerobic conditions [51,54]. Hg concentration in mountain firn core was found to be in the range of 2–35 ng·L^{-1} [55], and in freshwater from uncontaminated sites was in the range of 1–20 ng·L^{-1} [51], but a study by Moller and colleagues found in snow/brine concentrations of 70–80 ng·L^{-1} [53].

The effect of organic molecule pollution was addressed in a work by Cappa and colleagues [56], who investigated the impact of a summer ski resort on the structure of bacterial communities on a glacier surface. Their findings revealed that the concentrations of PAHs and PCBs were significantly higher close to the ski resort than in areas impacted only by long-range transport. They also observed that the presence of pollutants can favor the selection of bacterial strains able to metabolize them. In another study, after that an unexpected synthetic oil spilling happened in Taylor Valley (Antarctica) because of a helicopter crash, biodegradation was detected preferentially in sediment rather than in water and fluid-filled bubbles of an ice core [57].

These first evidences that organic pollution can select specific degrading microbial populations, thus promoting the biodegradation of the same contaminants on glaciers, were further confirmed by subsequent studies, which reported the actual biological removal under field conditions and used molecular cultivation-independent methods to better characterize the microbial communities. For instance, the study by Weiland-Brauer and colleagues found high concentrations of eighteen different congeners of PCBs, 16 PAHs and 29 different organochlorine pesticides in the cryoconite of an Alpine glacier [9]. In this study, microcosms containing cryoconite were set up with the addition of PCBs to determine the bacteria responsible of PCB degradation. The results showed that different genera, among which the most abundant were *Pseudomonas*, *Shigella*, *Polaromonas*, *Variovorax*, *Janthinobacterium*, *Subtercola*, and *Chitinophaga*, were able to degrade PCBs. On a molecular basis, the ability to aerobically biodegrade PCBs was identified in the presence of the gene *bph*A coding for the enzyme biphenyl dioxygenase, which was found in both metagenomic DNA and in the genome of the bacterial isolates.

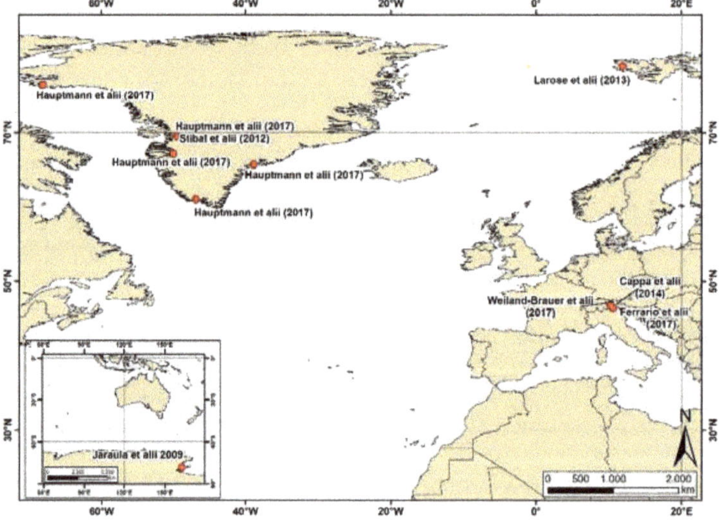

Figure 1. Geographic location of the studies addressing both the presence of pollutants and the microbial processes related to them in cryoconite and other supraglacial environmental matrices.

The microcosm approach was also used in two other studies with the aims of simulating glacier surface environments and estimating the biodegradation rates of pesticides [31,58]. In the first study, the ability of supraglacial bacteria to degrade the pesticide 2,4-dichlorophenoxyacetic acid (2,4-D) was investigated and, although no taxa related to known 2,4-D degraders were found, the authors succeeded in estimating the biodegradation rate in simulated field conditions. There are no data about the presence of this pesticide in/on glaciers, and not surprisingly, the rates of 2,4-D mineralization were lower than those observed in temperate environments. This work first demonstrated that anthropic contamination on glaciers can be attenuated by the local natural microbial community.

Ferrario and colleagues [16,31] detected high concentrations (2–3 mg kg^{-1}) of the pesticide chlorpyrifos (CPF) in the cryoconite and meltwater of an Alpine glacier, while in agriculture and termiticidal initial soils (i.e., in the application area) its concentrations vary from 10 to 1000 mg per kg of soil respectively [59]. The authors estimated, by in situ microcosms, that the rate of pollutant removal due to biodegradation was much higher than those of hydrolysis and photodegradation, leading to a CPF half-life in cryoconite ranging from 35 to 65 days (Figure 2). In other environments CPF half-life was reported to vary from 6 days (in aquatic aerobic conditions) to 128 days (in soil/slurry anaerobic conditions) [60], showing that in cryoconite the biodegradation rate is not different from others even if there are harsh conditions in this microhabitat. A whole shotgun metagenomics analysis of the bacteria occurring in the cryoconite of the same glacier allowed the genome reconstruction of a bacterial population that could be responsible of the pesticide biodegradation [31]. These putative CPF-degrading bacteria resulted to be photoheterotrophic *Burkholderiales*, thus strengthening previous findings on the metabolic versatility and the importance of Betaproteobacteria in the heterotrophic metabolisms of cold environments [37,61,62].

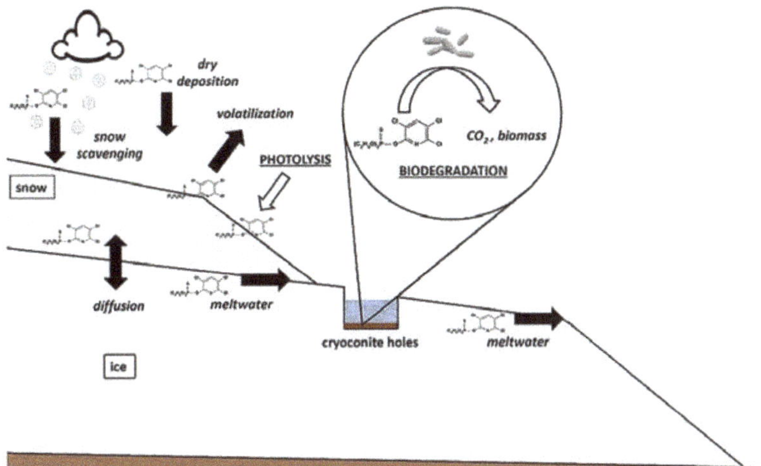

Figure 2. Processes affecting the fate of chlorpyrifos on the glacier surface. Biodegradation in cryoconite is highlighted. Reproduced with permission from Ferrario et al., 2017 [31].

A metagenomics approach was used also in a recent work addressing the metabolic potential of microbial communities in cryoconite collected on the Greenland ice sheet. Results disclosed the potential for resistance to and degradation of contaminants, including polychlorinated biphenyls (PCBs), polycyclic aromatic hydrocarbons (PAHs), and the heavy metals mercury and lead. The presence of genes for contaminant resistance and degradation was spatially variable but present in all samples across the ice sheet [63].

3. Conclusions and Future Perspectives

This short review of the studies that have addressed the biodegradation of pollutants on the glacier surface reveals how this field has been investigated only very recently, and that there is therefore a wide potential for future studies. New studies should investigate the biodegradation potential of glacier microbes for further pollutants that are known to occur on glaciers (e.g., terbuthylazine [16]). The review has also highlighted the importance of using microcosms to investigate degradation rates.

A well-designed experimental setup of microcosms under different conditions (e.g., light and dark conditions, sterilized and non-sterilized cryoconite, etc., see [16] for an example) can indeed allow for the assessing of the relative contribution of different processes to pollutant degradation. We also stress the importance of carrying out microcosm studies in situ, whenever this is logistically feasible, in order to be as close as possible to natural conditions.

Another important feature emerging from this review is the potential of metagenomic studies to improve our understanding of the metabolic processes that allow glacier bacteria to degrade pollutants. Indeed, the identification of the microbial populations involved in these processes was critical in most studies. With the decrease of costs of sequencing and the availability of new bioinformatic tools that will ease these studies, we hope for further investigations of pollutant degradation on glaciers combining microcosms and metagenomics.

We also hope that further studies will fill the large geographical gaps in these investigations, for instance in South America. Finally, there is also a strong need for models of the environmental fate of pollutants in cold areas, which incorporate also biological processes, as they will be pivotal to properly forecast the future dynamics of contaminants released by glaciers in future warmer climatic conditions.

Acknowledgments: The authors acknowledge the valuable contribution of all former researchers and students involved in the studies of microbiological processes in glaciers. Among them, they particularly thank Claudio Smiraglia, Ilario Tagliaferri and Claudia Ferrario.

Conflicts of Interest: The authors declare no conflict of interest.

References

1. Jaworowski, Z.; Bilkiewicz, J.; Dobosz, E.; Wódkiewicz, L. Stable and radioactive pollutants in a Scandinavian glacier. *Environ Pollut.* **1975**, *9*, 305–315. [CrossRef]
2. Sladen, W.J.L.; Menzie, C.M.; Reichel, W.L. DDT Residues in Adelie Penguins and a Crabeater Seal from Antarctica. *Nature* **1966**, *210*, 670–673. [CrossRef] [PubMed]
3. Ottar, B. The transfer of airborne pollutants to the Arctic region. *Atmos Environ.* **1981**, *15*, 1439–1445. [CrossRef]
4. Blais, J.M.; Schindler, D.W.; Muir, D.C.G.; Kimpe, L.E.; Donald, D.B.; Rosenberg, B. Accumulation of persistent organochlorine compounds in mountains of western Canada. *Nature* **1998**, *395*, 585–588. [CrossRef]
5. Kang, J.H.; Choi, S.D.; Park, H.; Baek, S.Y.; Hong, S.; Chang, Y.S. Atmospheric deposition of persistent organic pollutants to the East Rongbuk Glacier in the Himalayas. *Sci. Total Environ.* **2009**, *408*, 57–63. [CrossRef] [PubMed]
6. Hodson, A.J. Understanding the dynamics of black carbon and associated contaminants in glacial systems. *Wires Water* **2014**, *1*, 141–149. [CrossRef]
7. Mangano, M.C.; Sarà, G.; Corsolini, S. Monitoring of persistent organic pollutants in the polar regions: Knowledge gaps & gluts through evidence mapping. *Chemosphere* **2017**, *172*, 37–45. [PubMed]
8. Vanden Bilcke, C. The Stockholm Convention on Persistent Organic Pollutants. *Rev. Eur. Commun. Int. Envtl. L.* **2003**, *11*, 328–342. [CrossRef]
9. Weiland-Bräuer, N.; Fischer, M.A.; Schramm, K.W.; Schmitz, R.A. Polychlorinated Biphenyl (PCB)-Degrading Potential of Microbes Present in a Cryoconite of Jamtalferner Glacier. *Front. Microbiol.* **2017**, *8*. [CrossRef] [PubMed]
10. Borja, J.; Taleon, D.M.; Auresenia, J.; Gallardo, S. Polychlorinated biphenyls and their biodegradation. *Process Biochem.* **2005**, *40*, 1999–2013. [CrossRef]
11. Abraham, W.R.; Nogales, B.; Golyshin, P.N.; Pieper, D.H.; Timmis, K.N. Polychlorinated biphenyl-degrading microbial communities in soils and sediments. *Curr. Opin. Microbiol.* **2002**, *5*, 246–253. [CrossRef]
12. Cai, H.; Sheng, Q.Y.; He, Z.G.; Shi, W.L. Isolation, Identification and Characteristics of an Efficient PCBs-Degrading Strain. In *Advances in Energy and Environmental Materials*; Han, Y., Ed.; Springer: Singapore, 2017.
13. Calamari, D.; Bacci, E.; Focardi, S.; Gaggi, C.; Morosini, M.; Vighl, M. Role of Plant Biomass in the Global Environmental Partitioning of Chlorinated Hydrocarbons. *Environ. Sci. Technol.* **1991**, *25*, 1489–1495. [CrossRef]

14. Carrera, G.; Fernández, P.; Vilanova, R.M.; Grimalt, J.O. Persistent organic pollutants in snow from European high mountain areas. *Atmos. Environ.* **2001**, *35*, 245–254. [CrossRef]

15. Villa, S.; Vighi, M.; Maggi, V.; Finizio, A.; Bolzacchini, E. Historical Trends of Organochlorine Pesticides in an Alpine Glacier. *J. Atmos. Chem.* **2003**, *46*, 295–311. [CrossRef]

16. Ferrario, C.; Finizio, A.; Villa, S. Legacy and emerging contaminants in meltwater of three Alpine glaciers. *Sci. Total Environ.* **2017**, *574*, 350–357. [CrossRef] [PubMed]

17. Arellano, L.; Fernández, P.; Fonts, R.; Rose, N.L.; Nickus, U.; Thies, H.; Stuchlík, E.; Camarero, L.; Catalan, J.; Grimalt, J.O. Increasing and decreasing trends of the atmospheric deposition of organochlorine compounds in European remote areas during the last decade. *Atmos. Chem. Phys.* **2015**, *15*, 6069–6085. [CrossRef]

18. Villa, S.; Vighi, M.; Finizio, A. Theoretical and experimental evidences of medium range atmospheric transport processes of polycyclic musk fragrances. *Sci. Total Environ.* **2014**, *481*, 27–34. [CrossRef] [PubMed]

19. Grannas, A.M.; Bogdal, C.; Hageman, K.J.; Halsall, C.; Harner, T.; Hung, H.; Kallenborn, R.; Klán, P.; Klánová, J.; Macdonald, R.W.; et al. The role of the global cryosphere in the fate of organic contaminants. *Atmos. Chem. Phys.* **2013**, *13*, 3271–3305. [CrossRef]

20. Kallenborn, R. Persistent organic pollutants (POPs) as environmental risk factors in remote high-altitude ecosystems. *Ecotoxicol. Environ. Saf.* **2006**, *63*, 100–107. [CrossRef] [PubMed]

21. Grannas, A.M.; Jones, A.E.; Dibb, J.; Ammann, M.; Anastasio, C.; Beine, H.J.; Bergin, M.; Bottenheim, J.; Boxe, C.S.; Carver, G.; et al. An overview of snow photochemistry: Evidence, mechanisms and impacts. *Atmos. Chem. Phys.* **2007**, *7*, 4329–4373. [CrossRef]

22. Herbert, B.M.J.; Villa, S.; Halsall, C.J. Chemical interactions with snow: Understanding the behavior and fate of semi-volatile organic compounds in snow. *Ecotoxicol. Environ. Saf.* **2006**, *63*, 3–16. [CrossRef] [PubMed]

23. Morselli, M.; Semplice, M.; Villa, S.; Di Guardo, A. Evaluating the temporal variability of concentrations of POPs in a glacier-fed stream food chain using a combined modeling approach. *Sci. Total Environ.* **2014**, *493*. [CrossRef] [PubMed]

24. Bizzotto, E.C.; Villa, S.; Vighi, M. POP bioaccumulation in macroinvertebrates of alpine freshwater systems. *Environ. Pollut.* **2009**, *157*, 3192–3198. [CrossRef] [PubMed]

25. Schmid, P.; Bogdal, C.; Blüthgen, N.; Anselmetti, F.S.; Zwyssig, A.; Hungerbühler, K. The Missing Piece: Sediment Records in Remote Mountain Lakes Confirm Glaciers Being Secondary Sources of Persistent Organic Pollutants. *Environ. Sci. Technol.* **2011**, *45*, 203–208. [CrossRef] [PubMed]

26. Bogdal, C.; Nikolic, D.; Luthi, M.P.; Schenker, U.; Scheringer, M.; Hungerbuhler, K. Release of Legacy Pollutants from Melting Glaciers: Model Evidence and Conceptual Understanding. *Environ. Sci. Technol.* **2010**, *44*, 4063–4069. [CrossRef] [PubMed]

27. Margesin, R. Alpine microorganisms: Useful tools for low-temperature bioremediation. *J. Microbiol.* **2007**, *45*, 281–285. [PubMed]

28. Boetius, A.; Anesio, A.M.; Deming, J.W.; Mikucki, J.A.; Rapp, J.Z. Microbial ecology of the cryosphere: Sea ice and glacial habitats. *Nat. Rev. Microbiol.* **2015**, *13*, 677–690. [CrossRef]

29. Cook, J.; Edwards, A.; Takeuchi, N.; Irvine-Fynn, T. Cryoconite: The dark biological secret of the cryosphere. *Prog. Phys. Geogr.* **2016**, *40*, 66–111. [CrossRef]

30. Stibal, M.; Schostag, M.; Cameron, K.A.; Hansen, L.H.; Chandler, D.M.; Wadham, J.L.; Jacobsen, C.S. Different bulk and active bacterial communities in cryoconite from the margin and interior of the Greenland ice sheet. *Environ. Microbiol. Rep.* **2014**, *7*, 293–300. [CrossRef] [PubMed]

31. Ferrario, C.; Pittino, F.; Tagliaferri, I.; Gandolfi, I.; Bestetti, G.; Azzoni, R.S.; Diolaiuti, G.; Franzetti, A.; Ambrosini, R.; Villa, S. Bacteria contribute to pesticide degradation in cryoconite holes in an Alpine glacier. *Environ. Pollut.* **2017**, *230*, 919–926. [CrossRef] [PubMed]

32. Henry, S.; Bru, D.; Stres, B.; Hallet, S.; Philippot, L. Quantitative detection of the nosZ gene, encoding nitrous oxide reductase, and comparison of the abundances of 16S rRNA, narG, nirK, and nosZ genes in soils. *Appl. Environ. Microb.* **2006**, *72*, 5181–5189. [CrossRef] [PubMed]

33. Cook, J.; Edwards, A.; Takeuchi, N.; Irvine-Fynn, T. Cryoconite: The dark biological secret of the cryosphere. *Prog. Phys. Geogr.* **2015**, *40*, 66–111. [CrossRef]

34. Grzesiak, J.; Górniak, D.; Świątecki, A.; Aleksandrzak-Piekarczyk, T.; Szatraj, K.; Zdanowski, M.K. Microbial community development on the surface of Hans and Werenskiold Glaciers (Svalbard, Arctic): A comparison. *Extremophiles* **2015**, *19*, 885–897. [CrossRef] [PubMed]

35. Franzetti, A.; Navarra, F.; Tagliaferri, I.; Gandolfi, I.; Bestetti, G.; Minora, U.; Azzoni, R.S.; Diolaiuti, G.; Smiraglia, C.; Ambrosini, R. Potential sources of bacteria colonizing the cryoconite of an Alpine glacier. *PLoS ONE* **2017**, *12*, e0174786. [CrossRef] [PubMed]

36. Bagshaw, E.A.; Tranter, M.; Fountain, A.G.; Welch, K.; Basagic, H.J.; Lyons, W.B. Do Cryoconite Holes have the Potential to be Significant Sources of C, N, and P to Downstream Depauperate Ecosystems of Taylor Valley, Antarctica? *Arct. Antarct. Alp. Res.* **2013**, *45*, 440–454. [CrossRef]

37. Franzetti, A.; Tagliaferri, I.; Gandolfi, I.; Bestetti, G.; Minora, U.; Azzoni, R.S.; Diolaiuti, G.; Smiraglia, C.; Ambrosini, R. Light-dependent microbial metabolisms driving carbon fluxes on glacier surfaces. *ISME J.* **2016**, *10*, 2984–2988. [CrossRef] [PubMed]

38. Haritash, A.K.; Kaushik, C.P. Biodegradation aspects of Polycyclic Aromatic Hydrocarbons (PAHs): A review. *J. Hazard. Mater.* **2009**, *169*, 1–15. [CrossRef] [PubMed]

39. Kuppusamy, S.; Thavamani, P.; Megharaj, M.; Naidu, R. Biodegradation of polycyclic aromatic hydrocarbons (PAHs) by novel bacterial consortia tolerant to diverse physical settings - Assessments in liquid and slurry-phase systems. *Int. Biodeter. Biodegr.* **2016**, *108*, 149–157. [CrossRef]

40. Li, Q.; Kang, S.; Wang, N.; Li, Y.; Li, X.; Dong, Z.; Chen, P. Composition and sources of polycyclic aromatic hydrocarbons in cryoconites of the Tibetan Plateau glaciers. *Sci. Total Environ.* **2017**, *574*, 991–999. [CrossRef] [PubMed]

41. Dong, Z.; Qin, D.; Kang, S.; Liu, Y.; Li, Y.; Huang, J.; Qin, X. Individual particles of cryoconite deposited on the mountain glaciers of the Tibetan Plateau: Insights into chemical composition and sources. *Atmos. Environ.* **2016**, *138*, 114–124. [CrossRef]

42. Łokas, E.; Zaborska, A.; Kolicka, M.; Różycki, M.; Zawierucha, K. Accumulation of atmospheric radionuclides and heavy metals in cryoconite holes on an Arctic glacier. *Chemosphere* **2016**, *160*, 162–172. [CrossRef] [PubMed]

43. Baccolo, G.; Di Mauro, B.; Massabò, D.; Clemenza, M.; Nastasi, M.; Delmonte, B.; Prata, M.; Prati, P.; Previtali, E.; Maggi, V. Cryoconite as a temporary sink for anthropogenic species stored in glaciers. *Sci. Rep.* **2017**, *7*, 9623. [CrossRef] [PubMed]

44. Łokas, E.; Zawierucha, K.; Cwanek, A.; Szufa, K.; Gaca, P.; Mietelski, J.W.; Tomankiewicz, E. The sources of high airborne radioactivity in cryoconite holes from the Caucasus (Georgia). *Sci. Rep.* **2018**, *8*, 10802. [CrossRef] [PubMed]

45. Kim, C.S.; Lee, M.H.; Kim, C.K.; Kim, K.H. ^{90}Sr, ^{137}Cs, $^{239+240}$Pu and ^{238}Pu Concentrations in Surface Soils of Korea C. *J. Environ. Radioact.* **1998**, *40*, 75–88. [CrossRef]

46. Kato, H.; Onda, Y.; Teramage, M. Depth distribution of ^{137}Cs, ^{134}Cs, and ^{131}I in soil profile after Fukushima Dai-ichi Nuclear Power Plant Accident. *J. Environ. Radioact.* **2012**, *111*, 59–64. [CrossRef] [PubMed]

47. Shukla, A.; Parmar, P.; Saraf, M. Radiation, radionuclides and bacteria: An in-perspective review. *J. Environ. Radioact.* **2017**, *180*, 27–35. [CrossRef] [PubMed]

48. Hu, Q.H.; Weng, J.Q.; Wang, J.S. Sources of anthropogenic radionuclides in the environment: A review. *J. Environ. Radioact.* **2010**, *101*, 426–437. [CrossRef] [PubMed]

49. Margesin, R.; Zacke, G.; Schinner, F. Characterization of heterotrophic microorganisms in alpine glacier cryoconite. *Arct. Antarct. Alp. Res.* **2002**, *34*, 88–93. [CrossRef]

50. Larose, C.; Prestat, E.; Cecillon, S.; Berger, S.; Malandain, C.; Lyon, D.; Ferrari, C.; Schneider, D.; Dommergue, A.; Vogel, T.M. Interactions between Snow Chemistry, Mercury Inputs and Microbial Population Dynamics in an Arctic Snowpack. *PLoS ONE* **2013**, *8*, e79972. [CrossRef] [PubMed]

51. Morel, F.M.M.; Kraepiel, A.M.L.; Amyot, M. The Chemical Cycle and Bioaccumulation of Mercury. *Annu. Rev. Ecol. Syst.* **1998**, *29*, 543–566. [CrossRef]

52. Von Canstein, H.; Li, Y.; Leonhäuser, J.; Haase, E.; Felske, A.; Deckwer, W.D.; Wagner-Döbler, I. Spatially oscillating activity and microbial succession of mercury-reducing biofilms in a technical-scale bioremediation system. *Appl. Environ. Microb.* **2002**, *68*, 1938–1946. [CrossRef]

53. Møller, A.K.; Barkay, T.; Al-Soud, W.A.; Sørensen, S.J.; Skov, H.; Kroer, N. Diversity and characterization of mercury-resistant bacteria in snow, freshwater and sea-ice brine from the High Arctic. *FEMS Microbiol. Ecol.* **2011**, *75*, 390–401. [CrossRef] [PubMed]

54. Colombo, M.J.; Ha, J.; Reinfelder, J.R.; Barkay, T.; Yee, N. Anaerobic oxidation of Hg(0) and methylmercury formation by Desulfovibrio desulfuricans ND132. *Geochim. Cosmochim. Acta.* **2013**, *112*, 166–177. [CrossRef]

55. Wang, X.P.; Yao, T.D.; Wang, P.L.; Yang, W.; Tian, L.D. The recent deposition of persistent organic pollutants and mercury to the Dasuopu glacier, Mt. Xixiabangma, central Himalayas. *Sci. Total Environ.* **2008**, *394*, 134–143. [CrossRef] [PubMed]
56. Cappa, F.; Suciu, N.; Trevisan, M.; Ferrari, S.; Puglisi, E.; Cocconcelli, P.S. Bacterial diversity in a contaminated Alpine glacier as determined by culture-based and molecular approaches. *Sci. Total Environ.* **2014**, *497–498*, 50–59. [CrossRef] [PubMed]
57. Jaraula, C.M.B.; Kenig, F.; Doran, P.T.; Priscu, J.C.; Welch, K.A. Composition and Biodegradation of a Synthetic Oil Spilled on the Perennial Ice Cover of Lake Fryxell, Antarctica. *Environ. Sci. Technol.* **2009**, *43*, 2708–2713. [CrossRef] [PubMed]
58. Stibal, M.; Bælum, J.; Holben, W.E.; Sørensen, S.R.; Jensen, A.; Jacobsen, C.S. Microbial degradation of 2,4-dichlorophenoxyacetic acid on the Greenland ice sheet. *Appl. Environ. Microb.* **2012**, *78*, 5070–5076. [CrossRef] [PubMed]
59. Murray, R.T.; Von Stein, C.; Kennedy, I.R.; Sanchez-Bayo, F. Stability of chlorpyrifos for termiticidal control in six Australian soils. *J. Agric. Food Chem.* **2001**, *49*, 2844–2847. [CrossRef] [PubMed]
60. Tiwari, M.K.; Guha, S. Kinetics of biotransformation of chlorpyrifos in aqueous and soil slurry environments. *Water Res.* **2014**, *51*, 73–85. [CrossRef] [PubMed]
61. Darcy, J.L.; Lynch, R.C.; King, A.J.; Robeson, M.S.; Schmidt, S.K. Global Distribution of Polaromonas Phylotypes-Evidence for a Highly Successful Dispersal Capacity. *PLoS ONE* **2011**, *6*, e23742. [CrossRef] [PubMed]
62. Caliz, J.; Casamayor, E.O. Environmental controls and composition of anoxygenic photoheterotrophs in ultraoligotrophic high-altitude lakes (Central Pyrenees). *Env. Microbiol. Rep.* **2014**, *6*, 145–151. [CrossRef] [PubMed]
63. Hauptmann, A.L.; Sicheritz-Pontén, T.; Cameron, K.A.; Bælum, J.; Plichta, D.R.; Dalgaard, M.; Stibal, M. Contamination of the Arctic reflected in microbial metagenomes from the Greenland ice sheet. *Environ. Res. Lett.* **2017**, *12*, 074019. [CrossRef]

MDPI

St. Alban-Anlage 66

4052 Basel

Switzerland

Tel. +41 61 683 77 34

Fax +41 61 302 89 18

www.mdpi.com

Condensed Matter Editorial Office

E-mail: condensedmatter@mdpi.com

www.mdpi.com/journal/condensedmatter